PORTABLE COMMUNITIES

PORTABLE COMMUNITIES

The Social Dynamics of
Online and Mobile Connectedness

MARY CHAYKO

SUNY
PRESS

Cover artwork courtesy of Raul Villarreal.

Published by
State University of New York Press, Albany

© 2008 State University of New York

For information, contact State University of New York Press, Albany, NY
www.sunypress.edu

Production by Eileen Meehan
Marketing by Fran Keneston

Library of Congress Cataloging-in-Publication Data

Chayko, Mary, 1960–
    Portable communities : the social dynamics of online and mobile connectedness / Mary Chayko.
        p. cm.
    Includes bibliographical references and index.
    ISBN 978-0-7914-7599-7 (hardcover : alk. paper)
    ISBN 978-0-7914-7600-0 (pbk. : alk. paper)
    1. Interpersonal communication—Technological innovations—Social aspects.   2. Mobile communication systems—Social aspects.   3. Online social networks.   4. Interperson   II.   II. Title: Social dynamics of online and mobile connectedness.

HM1166.C48 2008
303.48'33—dc22   2008003115

10 9 8 7 6 5 4 3 2

For Mom, Daddy, John, and Cathy,
my very first community

He thinks of the many signals
flying in the air around him,
the syllables fluttering,
saying *please love me*,
from continent to continent
over the curve of the earth.

—Joseph Millar, "Telephone Repairman"

CONTENTS

I INTRODUCTION: THE INTERNET, MOBILE PHONES, AND COMMUNITY

1 **The Portability of Social Connectedness**   3

II INTERNAL DYNAMICS: INSIDE THE PORTABLE COMMUNITY

2 **Thinking in Tandem: Cognitive Connectedness**   17
   The Cognitive "Face" of the Community   18
   Sociomental Space   22
   Cognitive Resonance   25
   Stories and Collective Memories   31
   Proximity, Presence, and Reality   37

3 **Feeling Connected: Emotionality and Intimacy**   43
   Friendship and Intimacy   44
   Trust and Social Support   51
   Happiness and Hostility: The Moderation of Moods
      and Behavior   56

4 **Playing Around: Fun, Games, and Hanging Out**   63
   Games   64
   Just "Hanging Out"   69
   Playful Talk: Humor, Gossip, and Flirting   73
   The Seductive Allure of Fun   78

5 **Social Networking: Convenience, Practicality, and Sociability** 87
   Sociability   88
   Convenience   93
   Dating, Romance, and Sex   97
   Learning, Working, and Getting Things Done   101

III EXTERNAL DYNAMICS: THE PORTABLE COMMUNITY IN THE SOCIETY

6 Being There: Constant Availability   113
    Comfort and Companionship   114
    Emergencies   120
    Anxiety, Apprehension, and Overload   124
    The Impact on Privacy   130

**7 Harnessing Social Interaction: The Control of Time,
    Space, and People**   141
    Where, When, and Whether We Interact   142
    Technology-Based Strategies for Interaction   145
    Spontaneity and Social Interaction   155

**8 Creating, Expressing, and Extending the Self
    (and Watching Others Do So)**   159
    Socialization in a Technological Age   160
    The Making and Remaking of the Self   166
    Voyeurism, Watching, and Lurking   173
    Multitasking and the "Hyperlinking" of Identity   178

**9 Shaping the Social Landscape: Equalities, Inequalities,
    Possibilities**   183
    Technological Divides and Power Differentials   184
    Old Problems, New Angles   186
    Agency and Activism: Mobilizing for Social Change   191
    A Look Ahead   197

Acknowledgments   203
Appendix 1. The Methodology   205
Appendix 2. Profiles of Interview Subjects   215
Notes   233
References   255
Index   291

I

INTRODUCTION

The Internet, Mobile Phones, and Community

I

THE PORTABILITY
OF SOCIAL CONNECTEDNESS

"I love the feeling of being free and connected at the same time," a young woman confided to me via email late one night. "I can go about my business online, but still have access to all the relationships I maintain in person—family, friends, business. I see no downside to being this connected to my 'peoples.' "[1]

Although I have been researching online and mobile social connectedness for over ten years, people's willingness to share their feelings about the social connections they make technologically still sometimes strikes me as surprising. They tend to relate very personal feelings and experiences, often without ever having met me, in emails, instant messages, text messages. But, of course, as someone who researches this phenomenon, I shouldn't be surprised. My first book, *Connecting: How We Form Social Bonds and Communities in the Internet Age*, argues that it is both easy and common for us to form all kinds of social bonds and communities with people we have never met.[2] We each make hundreds of social connections with distant, even absent, others—connections that can be vivid, strong, reciprocal, and intimate.[3] In short, we form *real, consequential* social bonds with people we have never met face-to-face—and in this world of wireless computers and mobile devices we can do it nearly all the time, everywhere we go.

Since the publication of *Connecting*, people often tell me their stories of feeling bonded to distant, absent others—from faraway or dead family members to famous authors and historical (even fictional) figures, and, certainly, people on the other end of the radio, TV, telephone, or internet.[4] They often describe a strong sense of connection to these faraway others and punctuate their descriptions with comments like, "I've never told anybody about this" and "I thought I was the only one who felt this way!"—expressing a sense of relief, even catharsis, at the opportunity to talk of something usually kept private. But more often than not, after telling

3

me their stories, they pause, and add something to the effect of, "Don't you think it's sort of strange that I feel this way?"—which is really their way of saying, Are you sure I'm not just a little bit crazy?

Of course, they—*we*—are anything but crazy. Very little is stronger than our desire to form social bonds and groupings—a social culture—with one another. We routinely form connections with people from whom we are separated by space or even by time, and we will use almost any means at our disposal to do so. Print and electronic mass media, phones, computers, and all kinds of communication devices provide us with highly effective means of "getting to know one another," even across great distances. At the same time, these technologies assist us in expressing ourselves, in extending and revealing ourselves to one another, and in creating our societies.

We appropriate technology to create vibrant and complex social worlds that are very much a part of our lives. Online and mobile technologies are used to share thoughts, ideas, photos, music, audio, video—anything that can be transmitted technologically—in increasingly creative, sociable ways. These technologies are "bringing together the small contributions of millions of people and making them matter," says journalist Lev Grossman, who wrote the article in which *Time* magazine named the Person of the Year for 2006,

> You . . . [who] made Facebook profiles and Second Life avatars and reviewed books at Amazon and recorded podcasts . . . blogged about our candidates losing and wrote songs about getting dumped . . . camcordered bombing runs and built open-source software . . . who actually sits down after a long day at work and says . . . I'm going to blog about my state of mind or the state of the nation or the *steak-frites* at the new bistro down the street . . .[5]

Online and mobile technologies are now truly pervasive, assisting us in creating and sharing and connecting with others in previously unimaginable ways. And even those of us who do not (yet?) keep blogs, create avatars, and record podcasts and the like (definitions follow later in the chapter) are probably more enmeshed in the participatory nature of online life than we think. To live in a modern technological society is to use the technologies at one's disposal and with which one is comfortable (which may be as simple as using email, visiting websites, reading blogs, or using mobile phones) to express one's self and to reach out to others. In the process, we form social connections and bonds and networks and communities that can have real resonance and importance for us. More frequently, more easily, more *portably* than ever before, we form undeniable, if often subtle and invisible, social connections with one another.

As wireless technology has become more widely available and affordable, social connectedness has become, literally, untethered. Mobile and "smart" phones and devices, handheld personal data assistants (PDAs), MP3 players, and even notebook and handheld computers are small and lightweight enough to take with us nearly everywhere. Now, technology can be carried with us—even *on* us—all the time. It can accompany us in the car, the bathroom, the classroom, even in bed if we so desire (and, apparently, many of us do).[6] This means we now have access to hundreds, even thousands, of other people, at any time—to a whole host of groups and communities that are almost constantly available to us, even as we ourselves are on the move. Social bonds and communities are now easily made mobile and can be built, sustained, and accessed from practically anywhere at any time, or, in a word that I think covers all this more precisely and evocatively, they have become *portable*. This shift in the development and use of these technologies inspired what was to become my next big project, the one you will read about in this book—the *portability of social connectedness*.

As a sociologist with a background in communication and psychology, there is little here I do not think is interesting or important. The dynamics in and around these social connections are rich and distinctive and paint a colorful picture of modern life. Cognitive connections, emotionality, intimacy, playfulness, and social networking all emerge—often simultaneously—when social connectedness is technologically mediated. Shifts and changes in our behaviors and norms occur now at breakneck speed. And along the way, we—our selves, our relationships, our societies—are all changed. These dynamics of life in a society teeming with technological connectedness will be examined here, both theoretically and empirically, along with literature from the fields of sociology, communication, psychology, media and cultural studies, computer and information science, and many others; additional analysis from a number of experts in these fields; and my own original research—in particular, the 87 in-depth, open-ended electronic interviews conducted specifically for this book and referenced throughout (see appendix 1 for a more detailed discussion of the methodology). Though my source materials come from numerous and diverse domains, all but three of my interviewees are American; I would consider this, therefore, a study that best describes contemporary American life, with potential application to all technological societies.

"I blog, text, IM, email, and I don't like to ever be without my cell phone or have to shut it off—even in a theater," the woman referenced at the beginning of this chapter told me (I'm calling her SocialNetworking1— see appendix 2 for her profile and for those of all the individuals I interviewed). "Let's put it this way, my 'connections' are more important than whatever I'm doing that might force me to shut my cell phone off." The dozens of people who shared their thoughts and experiences with me will give shape and voice here to a timely set of issues—the construction, experience, and *meaning* of portable communities, to their members and to our societies.

※ ※ ※

Sociologists are experts in the study of life lived in social groupings. *Groups* are clusters or networks of individuals who share specific interests, ways of behaving, a common identity, and interpersonal interaction (among at least some members of the group).[7] They can be very small (even two or three people have been termed a group) or very large.[8] Whether this interaction takes place face-to-face or is mediated by some technology, something important, new, and almost indefinable happens when a group comes together: it develops an energy, a "charge," all its own—one that transcends, somehow, the sum of its parts. Sociologist Emile Durkheim writes of how a clan is "able to awaken within its members the idea that outside of them there exist forces which dominate them and at the same time sustain them" such that "a sort of electricity is formed by their collecting . . . (an) effervescence."[9] Groups "effervesce"—they have an "electricity," a power, that is all their own.

A *community* is a set of people who share a special kind of identity and culture and regular, patterned social interaction.[10] Ever since sociological theorist Ferdinand Tönnies declared community to be an essential condition for the development of close, primary social bonds (*Gemeinschaft*, which has become an enduring synonym for the traditional, indeed premodern, kind of community), sociologists have not been able to agree on how, or whether, definitions of community should be updated.[11] Some scholars even suggest that we discontinue our use of the concept altogether.[12] But I find it indispensable in describing many of the groupings that my interviewees (among others) identified and that will be examined here.

The very word "community" conjures up an image that matches quite well what many describe as their experience in online and mobile groupings. "Community" evokes a sense of neighborhood and neighborliness, of warmth and support and belonging, of close personal ties, of *Gemeinschaft* bonding. This idea is partly a misnomer, of course: not even the prototypical small town is *always* supportive and neighborly; its citizens are never *all* closely and warmly bonded. But it is a vivid, persistent image nonetheless, and useful for these purposes, for in examining online and mobile life, the first thing we must do is to make the invisible visible.

And if the small town metaphor is partly a misnomer, it is also more than a little accurate. There is much about the way that we form social ties and groups in the use of technology that is, for lack of a better word, communal. It is not quite the charmingly idealized small town, of course, but then, neither is that town itself—whether visited on foot or online. And yet technologically mediated communities are often much closer, more supportive, even more neighborly than might be supposed upon first glance. They generally consist of numerous criss-crossing *social networks*—sets of linked individuals whose patterns of connectedness form channels through which information, influence, emotional intensity, and sociability can flow (and even be measured and charted). Traveling these networks, members can derive *social capital*—resources and contacts that can enhance their status in society, sometimes dramatically. They can develop a *collective identity*, a specific, often strong, sense of themselves as a social unit. They can share a meaningful purpose and commitment. And within these social units, a wide range of social ties and linkages can be created. These can be primary or secondary in nature, local or extra-local, strong or weak (or something in between), and direct (that is, between people who actually contact one another) or indirect (between those who do not contact one another and whose knowledge of one another is only made possible via the mediation of some third party or technology).[13]

The use of such a visually accessible concept—the depiction of groupings of people connected by online and mobile technologies as actual *communities*, if technologically mediated and therefore potentially *portable*—gives us something we can all, mentally, "glance" at together, a common point of reference. Many find it an intuitively appropriate metaphor and use it casually to describe online and mobile life, as did nearly all of my interviewees:

> I feel I am part of a tight-knit community that cares about one another. (MusicLover1)

> IMing really feels like a community because everyone you know is online. (InstantMessaging4)

> You can definitely feel the community on the board and how it changes. (WorkGroup5)

> [My group] is an extremely tightly bonded community that simply cannot be found in normal daily life and that would do just about anything for each other—I sometimes describe my listserv as an electronic equivalent to the French Foreign Legion. (SupportGroup3)

Not a single one of my interview subjects questioned what I meant by "community": they used it comfortably, spontaneously, and similarly.

Portable communities, then, are groupings that use small, wireless, easily transportable technologies of communication (*portable technologies*) to facilitate interpersonal connectedness and to make and share a collective identity and culture. The term has also been used to refer to groups of people who physically move from place to place, assembling and reassembling periodically (as might groups of migrant workers, or fans who follow their favorite musicians to various concerts or festivals, in the manner of Grateful Dead fans or bluegrass fans and musicians).[14] This usage provides another apt metaphor for the kind of technologically generated communities we will examine here. Even when communities physically relocate, their spatially separated members must use some form of technology, whether portable (mobile phones and the internet) or more primitive (flyers or landline phones), to coordinate their efforts. In the portable communities examined in this book, technologies are used to bring people into contact with one another, though this may or may not eventually result in a face-to-face gathering. For our purposes, then, *portable communities* will refer to groups whose members connect via online and mobile technologies, whether they meet face-to-face frequently, occasionally, or never.

As with all kinds of groups, portable communities can be very small (a family, a group of friends) or very large (Benedict Anderson describes even the nation, or an entire culture, as a community)[15]—or, of course, something in between. They comprise the whole spectrum of online and mobile connectedness. Two-person units (called *dyads*) can regularly and easily email, instant message (IM), text message (text), and talk to one another via portable device. Small groups can communicate in online chat rooms or mobile text chat, discussion or message boards, in email or IM or text "round robins" or "text circles." They may connect on websites or *social networking sites* (websites specially constructed to help us connect through interactive profiles or pages we design and update—examples include MySpace and Facebook) and weblogs (most often now called *blogs*—online journals and/or sets of links that generally invite reader response and dialogue). Larger groupings of people can gather together on the most well-known blogs and websites including social networking sites, discussion boards and electronic mailing lists, and *wikis* (sites where content can be produced and changed by those who visit, like the online user-created encyclopedia Wikipedia). In these spaces, identities, sometimes represented by *avatars* and icons (depictions of a person that take the form of some kind of graphic) can be developed, and such media as photos, video, and audio (perhaps in the form of *vidcasts*, *vodcasts*, and *podcasts*—video or audio files that can be downloaded to a portable device) can be shared.[16] Portable communities are created when people use these kinds of technologies in any or all of these ways, separately or in combination, to develop a shared identity and culture.

In this book I focus less on particular qualities of each of these technologies—less on the differences among portable technologies—and more on the ways in which they, collectively, bring people together, wherever those people may be. Technologies such as computers, mobile phones, and PDAs (many of which double as phones, and may be called computer phones, smart devices, handheld computers, or some term yet to be popularized at this writing) are considered here similarly, and for the most part equivalently, as facilitators of portable social connectedness. For they are increasingly used in combination and in tandem with one another, and are even becoming interchangeable, performing multiple overlapping functions. Web applications can be accessed via mobile phones and PDAs, phone calls can be made via computer, photos and video can be taken with portable devices and easily exchanged.[17] And someday we will be connected by newer, smaller, even more ingenious technologies, yet to be mass produced, yet even to be invented, but certain to spark our collective interest, as technologies that facilitate an easy and portable sense of community tend to do.

Differences in the ways that these technologies may facilitate connectedness, then, are for the most part not relevant to this study. This book focuses more on the *experience* and *effects* of social connectedness as it is mediated by portable technology, and less on the attributes of the technologies and applications per se that bring it about. Therefore, when I refer to online and mobile technology in this book, I refer in a general sense to *any* and *all* of the technologies that can help us gather together in portable community. I will sometimes shorten the modifier to simply *online*, which increasingly refers to the accessing of web-based applications that facilitate emailing, instant messaging, text messaging, web surfing, blogging, chatting, photo-sharing, or any of a number of similar activities, from anywhere, using any technology that will do the job. I will sometimes refer to particular technologies and applications, but will more often reference the portability of social connectedness as it is facilitated by these technologies in general.

Even when we do something as simple as sending an email or text message, or talking to friends or family on a mobile phone, we can be establishing and strengthening community. This is because portable technologies tend to inspire strong user involvement. Online and mobile sites, requiring plenty of writing and reading (and sometimes speaking and moving about, as in internet radio shows, podcasts, vodcasts, and video sites like YouTube), enable and encourage interactive participation. As we access them, we can easily become involved: adding content, commenting, providing feedback or ratings, maybe even commenting on one another's comments. In the process, a network, a community, is formed. Sometimes such sites are built with the explicit objective of forming a network or community, as in an online class, an issue-oriented discussion board, or a support group. Sometimes the objective is more grand: to start a social movement, to influence a national debate, to set an agenda as to what is newsworthy. At other times, it is smaller, as when a social networking site or a blog is utilized to reach out to a very few others. But regardless of the scale and purpose of engaging in online and mobile behavior, when we use technology to interact and create a culture and a collective identity, online and mobile communities can spring up. Then, they can be linked to one another. It can be startling, in fact, how rapidly a sense of neighborhood and community can emerge, and how strongly people can feel about the experience of inhabiting these social spaces.[18]

Technologically mediated connections and communities are often referred to as *virtual*. When Howard Rheingold coined the term *virtual community* in his 1993 book of the same name, people hungry for an explanation of the then-new and rather mysterious phenomenon gobbled up his term *and* his thoughtful analysis.[19] Both have "stuck." Over time, "virtual" has become an almost universal descriptor for online phenomena, with one less-than-optimal consequence: "virtual" implies that whatever it describes is almost, or not quite, or "not really" real. It implies that something about it is illusory, imaginary, "less than."[20] Though subtle and often unintended, this is ultimately, I maintain, a damaging message. Online and mobile communities are absolutely and unequivocally real—as are many things (like radio airwaves, or love!) that can not be seen or touched. As I discuss further in chapter 2, the reality of something can not be measured by its physicality, but by the reality of its consequences (to paraphrase the famous sociological theorem of W. I. Thomas)—that is, the extent to which it has a real and genuine effect on something else.[21]

In my work, I favor the term *sociomental* over "virtual."[22] Any social exchange or environment in which people derive a sense of togetherness by being mentally oriented toward and engaged with one another can be described as sociomental. Two or more people must be involved in the exchange, which makes it *social*, and some degree of technological mediation is required to facilitate the connection and give us the opportunity to know of one another, which is the *mental* aspect. By using this term, we sidestep the misleading connotations of "virtual" and also describe more precisely the nature of a bond, community, or environment (in particular, those that are technologically mediated), placing the emphasis, appropriately, on the cognitive center or "core" of the relationship (for more on this idea as well, see chapter 2).

Sociomental connections and portable communities are sometimes (often, in fact) manifest in literal space. That is, they can be face-to-face as well as sociomental. There are *no* face-to-face communities (except the special, rare case of conjoined twins) in which all members are *continuously* in physical contact with one another. All communities, then, are sociomental at their "core." To be sure, some of them are "physicalized"—with at least some of their members meeting in physical space—from time to time. Such communities may be considered face-to-face as well as sociomental. But some communities are *purely sociomental.* Their members will never meet in physical space (elsewhere, I have described them as *communities of the mind*).[23] A group of people that encounter one another at least in part via portable technology, then, may become a portable community, regardless of the specific technology they may use (mobile phone, computer, PDA), the mode of communication or specific application employed (whether it be one-to-one or one-to-many; whether it occurs via texting, emailing, photo-sharing, electronic mailing list, etc.), and whether or not the people involved sometimes gather together in literal space. Indeed, it may be more useful to think of communities (and relationships) in general as existing along a continuum, with the sociomental at one end and the physically copresent (or face-to-face) at the other, and to use the continuum to consider the "degree of physicality" that the community or relationship may possess.[24]

Interestingly, research indicates that the use of online and mobile technologies tends to prompt, rather than hinder, face-to-face meetings. We often IM and text message one another to make dates to get together, email or talk on mobile phones to keep long-distance relationships viable, and gather on social networking spaces to stay updated on one another's doings. All of this makes impending face-to-face get-togethers more, rather than less, likely. More people use the internet to make new local connections, or to supplement existing connections between themselves and people they already know, than to engage in far-flung activities or global enterprise.[25] As sociologist Jeffrey Boase and his coauthors explain: "Contrary to fears that email would reduce other forms of contact, there is 'media multiplexity': the more contact by email the more phone and in-person contact. As a result, Americans are probably more in contact with members of their communities and social networks than before the advent of the Internet."[26] My interview subjects bear this out:

> With instant messaging I make plans to meet up with friends
> or decide what we will do that night, or catch up with
> friends and family that I may or may not talk with frequent-
> ly . . . (InstantMessaging1)

> With mobile connecting, you keep in touch with your friends
> and even if you are somewhere and you can't call them . . . the
> text message makes it easier to communicate with them. (Mu-
> sicLover2)

> I feel closer to my family and significant other due to constant
> mobile phone communication on a daily basis. I must speak to
> my bf at least 4-5 times a day and my kids usually once a day.
> I feel that due to this our relationships have stronger bonds, are
> deeper, and that we work out challenges quicker. The downside
> is I feel guilty when I turn the phone off . . . of course, this is
> very rare. (MobileUser1)

And then there is this funny (or sad, depending on your perspective)
story:

> My wife and I each have our own computers, since we've recently
> moved in together and used to live on our own. My computer
> is on the main floor and hers is in the basement. Both have
> internet access, so if we are each on our computers and I need
> to speak with her, I'll IM her rather than talk to her. It's mainly
> for two reasons: (1) I have a powerful computer . . . it's kind
> of loud, and (2) I don't feel like screaming. But even though I
> guess I have a legitimate reason for IM'ing my wife when she's
> just downstairs, I find it kind of pitiful. (TVFan1)

As we shall see, the ways in which we use these technologies are as varied
and diverse and intriguing as we are.

And the social ramifications of the use of portable technology are,
if possible, even more intriguing. New forms of social arrangement are
taking root: love affairs between people who might never have given one
another a second glance if they had first met offline; friendship circles in
which members do not even know one another's race or gender; groups
consisting of hundreds, even thousands of people who have never met but
who regularly and reliably provide one another with information, goods,
services, or heartfelt support. In what cofounder of *Wired* magazine Kevin
Kelly calls an "electricity of participation," portable technology

unleashes involvement and interactivity at levels once thought unfashionable or impossible. It transforms reading into navigating and enlarges small actions into powerful forces . . . [It] nudges ordinary folks to invest huge hunks of energy and time into making free encyclopedias, creating public tutorials for changing a flat tire, or cataloging the votes in the Senate. . . . The deep enthusiasm for making things, for interacting more deeply than just choosing options, is the great force not reckoned ten years ago.[27]

Though the web has supported interpersonal interaction since its inception, it would have been difficult to predict the kinds of participatory activities in which people now routinely engage and the kinds of linkages that are now possible. We can now create more, and more kinds of, shared culture than ever before, and can do so with people we have never met and may never meet.

As these connections and communities become more plentiful in our lives, it becomes ever more important to probe their social dynamics and implications. With this in mind, this book looks at portable communities from several different angles. It explores the *internal dynamics* of the communities themselves—the nature of the interactions within the groupings and the experiences people report as a consequence. These dynamics are cognitive (chapter 2), emotional (chapter 3), playful (chapter 4), practical, convenient, and sociable (chapter 5). The book also spotlights some of the *external dynamics* in effect—the ways that portable communities resonate with and in the larger society. It considers the impact on us, as individuals and as a society, when we are constantly available to one another (chapter 6), learn to control or "harness" our social interactions (chapter 7), discover new modes of self-development and expression (chapter 8), and grapple with the social problems and inequalities that result (chapter 9). It also takes a peek into the future, at a social landscape increasingly shaped by portable technologies (chapter 9). In sum, we examine here the experience and meaning of portable communities: how we create and sustain and are affected by them, sometimes in ways that threaten and hurt us, and sometimes in ways that help and heal us.

II

INTERNAL DYNAMICS
Inside the Portable Community

2

THINKING IN TANDEM
Cognitive Connectedness

Portable communities consist of people, places, and events—and the images of those people, places, and events that we carry with us in our heads. Colorful and complex, these are social worlds created largely of cognitive activity—of images we create and share of people, places, and activities. Because these "worlds" are cognitive, they are, logistically, portable. We can pick them up and take them with us wherever we go; we can access the others in them (certainly, at least, our images of them) at any time. And to an impressive degree we can coordinate our thinking *about* and *with* the others with whom we inhabit these communities.

This is a dynamic unique to the portable community. Families, friends, and lovers sharing physical space may become so close that they believe they know what another is thinking, almost reading one another's minds. But portable technologies allow us to do this nearly anywhere, anytime, and more reliably and satisfyingly than might be expected. We have long turned to technology to help us sustain our relationships: love letters have helped long-distance lovers stay close, phone calls keep family and friends updated on the latest news, video and audio recording allows us to see and hear one another in intimate detail. Technologies have always helped us to bridge distances, to maintain a sense of togetherness across space and time. But the technology of today provides us with portability, a powerful and effective tool in coordinating people's streams of thought. Now, we can *think in tandem* with people whom we may never have met but with whom we have much in common, and we can do this pretty much wherever and whenever we want.

In this chapter, we look at how this happens—how we can actually coordinate and synchronize our thoughts with absent others and how doing so builds and shapes our portable communities. The ways in which symbols, rituals, and temporal symmetry help us put a "face" on a

17

community that might otherwise be invisible will first be considered. The critical concepts of sociomental space and cognitive resonance, and their effects on community construction, will then be explored. We will examine the importance of stories and collective memories on the community, and how a sense of proximity and presence grants the community an undeniable "accent of reality." For amidst all this cognitive activity, a real social structure—potentially strong, solid, and enduring—can be built. This chapter brings to light this structure—what I call the *cognitive infrastructure* of a community and, collectively, of a society.

THE COGNITIVE "FACE" OF THE COMMUNITY

All communities are cognitive at the "core." Groups of people relatively rarely gather together—*all* together—in physical space. They are either too large or too widely dispersed, or their members too busy, for all of them to get together face-to-face more than occasionally (if indeed then). But that does not mean that they cease being "groupings" when they are not gathered. Groups and communities persist even in the dearth or absence of physicality, and even as members come and go.[1]

This is because *social connections and groupings are cognitive entities.* They are tied together by cognitive relations between and among their members. In fact, they exist in their most complete form *only* in the minds of their members.[2] Cognitive entities exist without the benefit of "face time." As such, they must be created and maintained without the impetus that face-to-face interaction can provide. So our minds must do a little something extra. Our minds must put a face to the objects of our connection, to our communities.

Symbols—visual or verbal or aural representations of other things—are indispensable for this purpose. They are found in all kinds of communities: sports teams and schools have their slogans, logos, and colors; friends and families have favorite games, nicknames, and catchphrases; religions and nations grant importance to icons, statues, pictures, documents. These symbols are endowed with great power; they cause us to feel things, often deeply and unconsciously. This power is derived from the power of the group; in fact, it *is* the power of the group. In his classic discussion of primitive religions and the importance of totemic emblems (or special symbols) to clans (or communities), Emile Durkheim explains how such power is derived and expressed:

> The idea of a thing and the idea of its symbol are closely united
> in our minds; the result is that the emotions provoked by

one extend contagiously to the other. . . . the symbol is something simple, definite, and easily representable, while the thing
itself, owing to its dimensions, the number of its parts and
the complexity of their arrangement, is difficult to hold in
the mind.[3]

The "thing" to which the symbols refer—which is often a clan or a
group—is aptly described here as an entity with a great number of dimensions and parts, an entity too complex to "hold in the mind." So,
Durkheim continues,

[The symbol] is treated as if it were this reality itself. . . . It is
therefore natural that the impressions aroused by the clan in
individual minds—impressions of dependence and of increased
vitality—should fix themselves to the idea of the totem rather
than that of the clan: for the clan is too complex a reality to
be represented clearly in all its complex unity . . .[4]

Symbols, then, serve as a kind of central vantage point for members of
communities to look to from their widespread settings. Since the group
is too complex (and often too large) to keep in our minds at all times,
the symbol brings the group into the minds of its members whenever it
is seen or perceived. As "by definition, it is common to all," the symbol
serves to bring a *similar image* of a group to the minds of its *physically
dispersed* members.[5]

This is how a community that is spread across space and even
throughout time can retain its "group-ness," its coherence of identity. We
look to the symbols of a community—its name, the photos on the website, the banner or logo that may be featured, the visual style of the page,
even linguistic style, the words and phrases and syntax that are used—and
immediately the community is brought to mind. And since these symbols
are essentially treated *as* the group, they have great power; they inspire
fervor and reverence. A Christian cross is not a mere piece of wood,
a flag is not something to casually stomp upon, and Yankee pinstripes are
far more than parallel lines.

Ritual activities performed by members of a community serve a
similar purpose. They bring the group to mind periodically and reliably.
When Muslims pray facing Mecca five times a day, a family gathers every
Thanksgiving for dinner, or an audience pauses for the national anthem
preceding a sporting event, a community is remembered and reinforced.
In portable communities we see all kinds of ritual engagements:

I have met one of my best friends online when we were virtually assigned as "due date" buddies. We email, IM, or talk on the phone every day! (ParentingForum2)

My kids keep in touch with me during the day, which does make my day pleasant. (WorkGroup6)

Email encourages people to check in on a regular basis. (OnlineInstructor)

A definite culture has come up around [my university's email system] . . . at college, I almost always leave it open and/or check it incredibly frequently. (SocialNetworking11)

Such acts help the group be "cognitively reunited" at regular intervals and ensure that even a very large group or culture that can never all gather physically (the set of all those who, say, have practiced a religion, followed a sport, gone to a school, or had any other meaningful characteristic in common) remains a viable social unit.[6]

The mass media provide another valuable assist in making portable communities visible and more easily brought to mind. Television, radio, newspapers, and magazines (and even old-school media like billboards, posters, bumper stickers, flyers, etc.) can popularize and spread a group's symbols, inspire rituals, keep communities in the public (and their members') eye, and even display actual members of the community from time to time. These members then become additional symbols that "stand in" for the community as a whole. Those people we view in the media—participating in a group or perhaps fighting for a cause—are perceived by us as *typified others*, more or less typical representatives of the group as a whole.[7]

People who are fans of, for example, a television show or musical group, or who may belong to a political party or activist community, can see other members of the group in action at various events: cheering at a televised convention, marching in a parade, discussing issues on a radio program or in an online discussion forum. They may see symbols of the community publicly displayed on stickers, shirts, or flags. They may then find a way to participate more actively in the community, even from their distant location—producing a blog or a podcast, starting a website or a text circle (in which a number people email or text-message one another "around and around"), gathering in a chat room or discussion board, participating in an online gaming world. Any of these activities, especially when regularly and ritually practiced, can strengthen the group and help bring it more steadily and concretely into members' minds. People create and access the image of the group, and build and maintain their bonds with one another, through these kinds of ritualized "cognitive reunions," even if they do not often or *ever* meet up physically with the others in the community.[8]

Symbols and rituals, then, often media-assisted, make communities more "visible" and help them to be brought similarly into their members' minds. Communication and game researchers Lisbeth Klastrup and Susana Tosca tell us that game players and designers "share a mental image" of the massive, complex cyberworlds they create together.[9] These "worlds" or communities are jointly elaborated and jointly changed over time. As their members create and share similar mental images, communities are given shape and form, given an all-important "face." And group members themselves are given faces as well. When we think about others, we orient ourselves, cognitively, to specific images of them—generally, to their faces. If no face is available, we create one in our minds, along with a personality, just as WorkGroup5 does:

> People are known by their handle or username and the image associated with it. In your own head you always make up an idea of what someone looks like and what their personality is just from what they post.

It is hard to bond to an idea; we bond with people, and if we have no visual point of reference we form pictures in our minds of what we imagine they may look like.

Sometimes we seek evidence for the veracity of our mental images, while sometimes we prefer to leave it to our imaginations. The fact that the created image may not be accurate seems less important to us than simply having one. We can actually bond more strongly to someone when we do *not* have a visual image provided for us than when we *do*, because when there is no image provided, our minds must work overtime to create and sustain an image, devoting extra energy to the getting-to-know-you process. Filling in the missing details is a highly involving cognitive task that can result in an even tighter interpersonal bond. "People can learn quite a lot about each other in the process of writing," sociologist Andrea Baker points out. "Their writings can communicate their strengths and weaknesses, as well as their worldviews, types of humor, and what they seek in another person."[10] With no details provided other than those that we gather from words, we can create mental images to which we orient ourselves and become bonded.

This is ultimately what makes online and mobile communities portable. The images and symbols and memories of particular and ritual events are kept with us, even "in us," all the time. Firmly fixed in our minds, ready to be activated upon the slightest reminder, the community is, in a very real, concrete way, brought with us wherever we go. As SocialNetworking5 puts it:

> i get to keep up with friends from high school that i don't get to see very often anymore. by seeing the groups they are involved in, by their interests and hobbies, i can keep up with them and still feel as if I am connected to them.

"Seeing" the groups in which her friends are involved helps SocialNetworking5 sustain her connectedness to them. It helps her feel that her friends remain very much a part of her life.

We use the technologies at our disposal, whether they be cameras, old-fashioned letters, the print and electronic media, or computers and mobile devices, to bring our friends, coworkers, and members of other communities to mind. In addition, we seem to need to imagine the community as situated in some kind of space. An understanding of the process by which we cognitively create such spaces is critical to understanding portability in social connectedness.

SOCIOMENTAL SPACE

Human beings need to use spatial imagery in order to organize and envision things cognitively.[11] Indeed, "portable community" is a metaphor that implies spatiality; it encourages comparison to space-based groupings. But, of course, we can not share physical space when we are physically separated from one another. We share a kind of cognitive space that I call *sociomental space*.

Sociomental space is the cognitive analog to physical space. It is a kind of mental habitat where portable communities "gather" and where cyberspace can be said to be situated. As writer and cyber-futurist Bruce Sterling describes it:

> Cyberspace is the "place" where a telephone conversation appears to occur. Not inside your actual phone. . . . Not inside the other person's phone. . . . The place between the phones. The indefinite place out there, where the two of you, two human beings, actually meet and communicate . . . cyberspace is a genuine place. Things happen there that have very genuine consequences.[12]

Sterling's characterization of cyberspace is beautifully illustrative, especially as he makes the turn from telephony to computerization:

> In the past twenty years, this electrical "space," which was once thin and dark and one-dimensional—little more than a narrow speaking tube, stretching from phone to phone—has flung itself open like a gigantic jack-in-the-box. Light has flooded upon it, the eerie light of the glowing computer screen. This dark electric netherworld has become a vast flowing electronic landscape.[13]

This space has become, in cyber-political scholar Tim Jordan's words, a "social, cultural, economic and political space" filled with human interaction.[14]

In order to accurately convey the nature of activity "in" such a space (and ensure that such activity is seen as truly authentic) this space needs to be carefully theorized. I find it useful to consider sociomental space as a "place" in which groups of people who may be physically separated create connections, bonds, communities, and entire social worlds. To do this, *mental maps* of these social worlds are created, then constantly used and updated.

We need a method to store and reference the mental images we are constantly generating. They are encoded and stored in the form of mental maps. We all have many "maps" in our heads that help us find our way in the world. These important tools help us collect, organize, and use information about our whole social environment, including the people in it and the places we go. Mental maps are multisensory topological representations of our physical and social worlds, consisting not only of mental images but sounds, smells, tastes, and feelings. We use them, for example, to situate in our minds whoever is above or below us in a hierarchy, who is warm or cool toward us emotionally, and with whom we feel close or distant. Each of us has our own individualized set of mental maps that organize and circumscribe our social environments, and we keep them in our heads to guide us wherever we go.[15]

When two or more people form a social connection, they go on to construct a collective or group map that represents their "common ground"—their shared understandings and shared environment. It is as though the individual mental maps of each of the people involved have overlapped and intersected and resulted a new social product that could not have been created in any one mind but requires the collaboration and creativity of a bunch of minds. These group maps represent the space that group members collectively inhabit; they are where "meetings of the minds" take place. And I'll go one step further and suggest that the set of all the individual and group mental maps in a society can be considered the sociomental space of that society.

With this construct, it becomes easier to envision cyberspace as a special subset of sociomental space—one created by groups of people who use cyber (online *and* mobile) technologies together, creating whole social worlds together. *Cyberspace*, in this view, is the set of all the individual and group mental maps created in cyber-communities; a real, if mental, place created in the joint delineation of hierarchies and networks and social bonds by members of such groups. It is created in the collective stream of thoughts, responses, information, feelings, and rushes of energy that are endlessly exchanged whenever people come together online. As InfoGathering1 told me:

> This is *where* I speak to friends I have not seen in a while, *where* we keep each other posted on what's going on in our lives, or just talk nonsense; *it's our little gathering place.* (emphasis added)

Like all sociomental spaces, cyberspace is multi-dimensional and multi-sensory, complete with mentally imaged topography, hot and cold spots, colors and textures.

We borrow the imagery and metaphors of physical space in creating these mental maps. To organize cyberspace, we are accustomed to speaking of "windows," "files," and "folders" on "desktops." Gamer1 described one favorite site as "my cabinet of curiosities"; SupportGroup2 described another as "a window to the world." And MusicLover1 reports that:

> The most important function of this forum is it is a *place* where rock fans can *gather* to celebrate the band they love and talk about everything music! It is a place where we can all express our passion for our favorite band! (emphasis added)

Or as SupportGroup5 says:

> A blog is where I want to be.

We use spatial metaphors freely in describing mental and sociomental activity because we can hardly do otherwise. We are traveling what seems to us to be *terrain*. To understand what we are doing and where we are going, we map this terrain, individually and collectively. We fix these sociomental spaces as securely as we can in our brains so that we have a system always at hand to help us negotiate our increasingly complex interaction-filled online and mobile experience.

According to communication theorist Joshua Meyrowitz, the mass electronic media have provided many millions of us with a common set of images and experiences. Even as physical place and location have become less critical to so much of what we do, he says, "electronic media move people informationally to the same *place*" (emphasis added).[16] Though his classic analysis of media and social place predates for the most part the internet and mobile technology era, there are clear analogies to the ways that portable technologies now provide people with a common sense of place. "Wherever one is now—at home, at work, or in a car," Meyrowitz noted in 1984, more presciently than he probably could have imagined, "one may be in touch and tuned-in."[17]

Subtly, constantly, mental maps help us think about, talk about, understand, and manage all the social connections we make. They help us envision and locate (and, most importantly, envision and locate *similarly*) the sociomental spaces and cyberspace places that are so much a part of all of our lives. As we use them, we situate the social connections and communities we make, both face-to-face and at a distance, and the places we go, both physically and mentally. Of course, some of these connections are more important to us than others. These result in the development of *cognitive resonance* in our relationships.

COGNITIVE RESONANCE

With a cognitive face, in a sociomental space, we come together in online and mobile communities. Some of us even become so close in them, so intimate, that we can be said to truly "think in tandem." This happens when we find people whom we understand very well, and with whom we feel that we are understood in turn. With these (relatively fewer) people, cognitive resonance can be generated.

The precondition for cognitive resonance is the establishment of *intersubjectivity*. Although we do not have direct access to the "inner lives" of others, we can still come to understand them, or, at least, to believe that we do.[18] We do this by assuming that others think about things more or less as we do unless we are presented with specific evidence to the contrary. We do this unconsciously, effortlessly. It would be extremely disruptive to everyday interaction if we were to continually halt the process and undergo in-depth investigation into whether others really understand us as we think they do, and we them. Instead, we tend to—we want to—assume understanding unless given reason to believe otherwise. Intersubjectivity, assuming and then building these common understandings, allows us to build up a common stock of knowledge, to find common ground. "All collective actions are built on common ground and its accumulation," explain Herb Clark and Susan Brennan, the developers of common ground theory.[19]

Our social worlds are built on the presumption and creation of common ground. Of course, we can think differently from someone else in many ways and still create a community or social world together. But there is generally some underlying commonality, even if there are many differences. Just as we are drawn to those whom we understand, we are also attracted to difference, and to *complementary differentiation*—differences that complement and "balance out" our own traits (as when an extrovert and an introvert, or a cautious person and a risk-taker, form a connection). This kind of connection may be more challenging than one characterized by a greater amount of similarity, but it can be just as strong.[20] And few relationships are either/or—generally people have qualities in common *and* those on which they differ.

When people establish common ground or share a social world together, they may find that they "resonate" in a special way with certain others. *Cognitive resonance* is the kind of interpersonal closeness we talk about when we say that certain people "click"—that they enjoy a special "spark" or "vibe," or are "on the same wavelength." This kind of connection, again, can be rooted in similarity, complementary differentiation, or some blend of the two. Andrea Baker, who studied 89 couples that met and found romance online, calls this the "simpatico feeling . . . the indefinable something, the 'click' between two people, the chemistry . . ." Or as one of the people she surveyed put it:

> The moment Bud entered the [online] channel I was aware that
> he was different and special. There is no explaining this.

The almost mysterious dynamic of cognitive resonance—"that x-factor, the elusive chemistry that attracts two people" that Baker describes—is very much in evidence in online and mobile communities and indeed can be propelled by connecting at a distance. And it can, Baker confirms, "be present from the start."[21]

For sociological theorist Alfred Schutz and communication scholar Wilbur Schramm, this kind of resonance occurs when people not only come to understand one another in a special way but *communicate* this sense of understanding and being understood to one another.[22] Schutz called this the "tuning in" process; it is how what he called *we feeling* is created, a sense that a collection of people has become a unit, a "we." We all need and want to be understood, and to understand others in return. We look to see something of ourselves in the other—a determination we sometimes make correctly and sometimes incorrectly, and often make prematurely. But finding someone whom we feel understands us in a meaningful way—our needs, interests, desires—is an extremely powerful, resonant thing. In general, the more people believe they are similar to another person, the more they like the other and the more they disclose about themselves.[23] But, of course, opposites attract as well.

Portable technology provides us with the means to communicate our thoughts and feelings to many with whom we may have something in common, and therefore increases our opportunities to find cognitive resonance with others. My interviewee E-dieter thinks of it this way:

> The internet is a central meeting place for people of the same characteristics, interests, and so on, no matter how specific, unusual, or taboo. If a person wants to find a tattooed one-eyed lesbian albino who likes to play basketball and watch Seinfeld, they will probably have a better chance finding the person on the internet than in a local bar.

Mobile phones, other PDAs and wireless devices, and computers help people who may have something in common find one another and get to know one another. This can be done with the explicit goal of finding someone with whom we might "click," as on a dating or matchmaking social networking site, or it can happen serendipitously, in the course of doing other things online.

As are most communities, portable communities are generally founded on common interests, goals, and values.[24] As we begin to share thoughts, ideas, and information, we may find we want to get to know more about (some of) the people with whom we are in contact:

> I am part of MANY online groups. I visit every one of them on a daily basis and interact on all . . . I get to interact with people who have the same interests as I do and get introduced to some things that they like as well. (MusicLover3)

> I feel there is an anonymous kinship because we feel the same about the shows we like. It's exciting to find others half way across the world who hold the same thoughts and interests as you do. (TVFan2)

> I like to talk to people who are like minded and make a site for myself . . . it is a fun way to talk to keep in touch with

other "otaku" (anime geeks) and to learn more about newer shows. . . . It makes me feel happy to talk about my show. There's just something about finding a show you both like. It's hard to stop talking about it. (AnimeLover)

As we get to know more and more about one another, some of us are likely to discover that we are "like-minded." When communities are technologically portable and can be visited frequently, the likelihood that we will encounter like-minded others, and become close with some of them, is dramatically increased.

Online or offline (that is, face-to-face), the establishment of cognitive resonance can have a profound effect on us. Interpersonal "chemistry" is actually the result of neurochemicals like dopamine and norepinephrine flooding the brain's pleasure centers, similar to the way that cocaine or chocolate or gambling can stimulate them. That is, cognitive resonance can create a "high," one that we want repeated and repeated again, one that makes it difficult to think rationally or even to concentrate on anything else. Detecting strong resonance or "chemistry" with another person taps into some of our most profound human forces—our cravings for new and novel experiences, for interpersonal closeness, and to simply feel that we are palpably, even wildly, alive. "We all seek that intensity," argues noted drug addiction researcher Nora Volkow. "There's something very powerful about that."[25]

When two people come to resonate strongly with one another, they begin to affect one another so powerfully that their brains are altered. They become able to deduce one another's thoughts and intuit one another's feelings. Their brains become interlinked. This is how we literally begin to *think in tandem* with one another. Our brains, our behaviors, even our bodies can become literally reshaped as a consequence of intimate connectedness.[26] Inter-brain neural linkages occur in all types of interpersonal interactions, even those that are fleeting, but are especially pronounced when people become very close.

When we form close relationships, our thoughts and actions can become synchronized. This can take the form of our unconscious mimicking of another's facial expressions and movements or of synchronizing our speech patterns to another's. Online, a kind of textual synchronicity can develop whereby people gravitate toward a similar syntax. In human connectedness, as with many other social and natural phenomena, we constantly adjust and entrain to one another's rhythms and patterns, often in tacit ways. This helps us to be brought experientially, if not physically, "together," and become firmly "linked" even if we are rarely, or never, physically copresent to one another.[27]

Synchronicity and togetherness are also created when people who may be spatially separated focus on the same things, in much the same way, at the same time. This can occur in situations as different as enjoying a work of art or music together, watching a TV show or listening to a radio show together, using a telephone sex or party line, rallying around flags, singing anthems, celebrating holidays or experiencing episodes of nationalism, and following schedules and calendars together.[28] Synchronous online and mobile activities can provide the same kinds of satisfactions. In experiencing things at the same time, possibly in the same way, our internal rhythms, intuitions, and mentalities become, effectively, synchronized. It provides the physically separated with an opportunity to go through an experience *together*. Sociologist Eviatar Zerubavel calls this *temporal symmetry*—the timed, coordinated sharing of action and thought.[29] When something that has meaning to them happens to spatially separated people at the exact same moment, they can come to feel that they are sharing time and experience in a particularly resonant way. This happens when spatially separated people watch the ball drop on New Year's Eve at the exact same moment (midnight) on television, see a message pop up that indicates that a favorite friend is online at the same time they are, or even physically peer into another car at a stop light to discover that the person in it is enjoying the same song on the radio. Technology brings people together in ways that can seem mundane but in actuality feel quite special.

Portable technologies are prime facilitators of temporal symmetry. They help us find one another easily and often, no matter where we are, even if we are physically on the move. When brought cognitively together, the thoughts, actions, and energies of disparate individuals can almost seem to merge. We come to share a sector of sociomental space (or of cyberspace) and can develop what sociologist Karen Cerulo calls a "sense of cognitive cohesion."[30] John Perry Barlow, one of the first social thinkers to conceptualize community on the internet, says that when online

> I want to be able to completely interact with the consciousness that's trying to communicate with mine. Rapidly. . . . We are now creating a space in which the people of the planet can have that kind of communication relationship.[31]

Temporal symmetry via portable technology makes possible the kind of rapid interaction with someone else's "consciousness" of which Barlow speaks. Synchronous online and mobile activities such as IMing, texting, live chat, and "live blogging" events provide numerous such bonding opportunities. As ParentingForum 2 says:

> I visit two boards for children born around the same times as mine were born, and it is nice sometimes to see that other moms are experiencing the same things I am at the same times.

Because they enable collective thought and emotion so well, portable technologies are perfectly positioned to engender social synchronicity, cognitive resonance, and thinking in tandem.

As we detect cognitive resonance with some of the others in our communities, close personal relationships spring up:

> I think I know the personal side of some of these people more than I do my own family members. Why, because people with cancer often let their "hair down" (that is if they have any—ha ha) easily when they are connecting with others who faced similar battles. The witty gallows or dark humor can really be quite hysterical. (SupportGroup1)

When we feel that we have gone through similar experiences and may even be similar to others, it speaks to our deeply rooted need to be understood, accepted, validated. These needs may be so deep that we are neither consciously aware of them nor realize that they may motivate us to take part in portable communities:

> I just love the feeling of being able to go on and read what other people are writing, especially on topics that I love, such as religion, race, gender, culture, and politics . . . there are usually fifty different people describing one topic and each have different opinions and feelings. I hardly chat myself, even if others address me, due to the language being used sometimes, but *I just like being there and I don't know why*. (InfoGathering3; emphasis added)

Even as InfoGathering3 feels drawn to her community, she has difficulty describing its specific appeal, and perhaps has never been given "permission" to explore such feelings. We don't always know why we are attracted to certain others. And as there is still sometimes a stigma attached to online connecting, especially when entered into enthusiastically, it may seem a little strange or shameful, and the attractions, bonds, and relationships we form there may seem odd or ill-considered, somehow "out of bounds."[32] Of course, as I argue throughout all of my work, online and mobile connections are commonly and frequently formed, and are part of the lives of many, many millions of people. But they are still too often hidden away, enjoyed in secret, as though they are blights on a portfolio of social connections instead of a fairly ordinary component, which hinders our understanding of them.

To discover that we think similarly to, and may even *be* like, certain others is healthy and heartening in the fullest sense of those words. Some of these others will be known to us face-to-face, others will not. To feel a strong sense of cognitive resonance with anyone else is naturally exciting; it trips a number of cognitive pleasure circuits. To be able to access those that inspire these kinds of feelings in us most any time, nearly everywhere we go, can be downright heady. Some even say that connecting frequently with others can provide an antidote to the stresses and alienation of modern life.[33] It is no wonder, then, that we find such a rich source of collective resonance in online and mobile communities, and that we begin to build up a bank of collective memories and stories.

STORIES AND COLLECTIVE MEMORIES

Every community, like every person, has a story. As individuals and as groups, we are prone to constructing narratives for our lives; we like to view past events and project the future as episodes in an overarching story. To bring order to seemingly unconnected series of events, to bring structure to our lives, we construct tales: of glory, of defeat, of progress, of togetherness. As Eviatar Zerubavel explains in his study of the social shape of the past, stories make past events meaningful and coherent and give our lives context. When we write our resumes, for example, we "present our earlier experiences and accomplishments as somehow prefiguring what we are currently doing."[34] This makes our lives seem more orderly than they actually are. We impose orderliness onto life through the telling of stories.

We do much of this storytelling in online and mobile communities. We construct narratives of threaded posts, debates, the profiles and photos and avatars that represent group members—of all our exchanges. These stories take into account the groups' histories; indeed, the history becomes the story (Zerubavel points out that the French and Spanish languages have a single word for "story" and "history.")[35] The telling and retelling of the story of any group is a common communal ritual: it gives a group definition and cohesion and creates solidarity among members. And, of course, the rise of the web as a participatory technology has resulted in an avalanche of storytelling online, in social networking spaces, blogs, podcasts, vodcasts, wikis, and any number of websites.[36]

Stories are highly evocative of person and place. They help us create and sustain the cognitive face of the group, envision the sociomental space in which its members reside, and detect cognitive resonance with one another—providing much-needed detail to the cognitive images we create. Telling one another our stories reveals us as "whole persons" and explains why we do what we do—or why we *think* we do what we do—or, at the very least, why we *say* we do what we do.[37] Just as a cohesiveness and order is brought to seemingly scattered, unconnected messages and ideas in storytelling, so can cohesiveness be brought to our own identities as we give them text and voice and then share them with others.

Online and mobile technologies help foster our love and need for narrative. Personal blogs fill this bill nicely, providing platforms for us to create a story of a series of postings, photos, even videos. Often regularly updated, blogs permit a community of readers, large or small, to react to, comment on, and in effect add to the blogger's story. Fully half of all bloggers blog just for this reason: to document their own experiences and share them with others.[38] They are, according to Amanda Lenhart and Susannah Fox of the Pew Internet and American Life center for research, "the internet's new storytellers."[39] People also tell their stories and share their experiences on discussion boards, websites, electronic mailing lists, and all kinds of portable communities.

These stories provide us with all-important images with which to populate our sociomental spaces. When I teach an online class, for example, I generally have students spend some time on the first discussion board telling stories that reveal their interests and personalities. This helps give the participants "faces" that are similarly, collectively, "seen." The class, then, is more easily envisioned as a whole, which helps it cohere and, ideally, develop into more than a narrowly focused work group. The goal is that it will become a true community whose members assist one another in learning, much as they do in more traditionally formatted (successful, that is) classes. Stories serve to "circumscribe time and space . . . and elevate the personal over the institutional."[40] The more detailed and resonant the story, the more "personal" and less "institutional" these spaces can be—that is, more "communal"—and the more vividly the people in them can come to life.

Using a variety of portable technologies can expand one's ability to tell a story. On a podcast, one can express himself or herself through voice and words. Webcams and video- and photo-sharing can lend a visual component to the narrative. Camera phones are increasingly used to tell stories; they are sometimes even used as personal photo archives that essentially "contain" the story of a person's day or week (or life?). They are now often used to take numerous quick, casual shots and videos of all kinds of things once rarely thought worthy of film (scenes from a daily jog, last night's dinner, even the purchase of new pair of shoes!). When saved or shared, these can form a narrative, especially if posted on a blog or online photo-sharing site with commentary that others can contribute to. They represent a very detailed kind of "personal self-authoring"—an ongoing, ad hoc visual archive of one's life.[41]

Sometimes, the narratives we create are fictional, as in *fanfiction*—scripts and stories written and shared by fans of a particular media offering or genre:

> I like to read fanfiction of anime online. It is a fun way to continue on with a series when you are done watching all the episodes, when you just can't stop thinking about it. . . . It's just a rush I can't let go of, though I have tried. (AnimeLover)

The "rush" AnimeLover speaks of reflects the pleasure of participating in an ongoing (in a sense serialized) narrative *and* of resonating with others along the way. As we shall see in the next chapter, spending time in online and mobile activities can be highly emotional and often results in this kind of "rush."

Stories serve as a platform for us to explain things to ourselves and understand our social worlds. As such, they help give something lasting meaning.[42] And many members of portable communities take the telling of their stories and those of the community quite seriously. They tell one another quite a bit about themselves, roaming well beyond the official boundary or focus of the group to get to know one another in depth. The history of the group itself is often shared as well. This may happen as new members are welcomed, when long-standing members are ritually reunited, as in periodic "roll calls" in which members each post and say something about themselves, or at special online or offline events. Telling stories about themselves and about the group promotes thoughts and feelings of togetherness, in addition to giving the group a "face."[43] Stories draw us in to the community and to one another's lives; they tell us that our experiences are indeed shared. In sum, they become the collective memory of the group.

Stories help dispersed members feel that they are in unity, and in celebrating and remembering the past, to look to a shared future. As events are interpreted through stories, the meaning of these events or phenomena can be better understood and the collective memory of the group expressed.[44] This facilitates the building up of collective memory and group identity. In fact, the memory storage and retrieval system in the human brain is quite flawed and unreliable compared to technological systems of maintaining and retrieving communal memory. Technology can be preferable to, and more effective than, face-to-face means of assembling and sharing collective memory.[45]

Online, collective memory exists formally, in archived or written form, and less formally, in the memories and exchanges of long-standing members.[46] Computer science and information technology researcher Gary Burnett reminds us that any text-based community

> is defined by its own particular history of writing and reading—a history documented by an ever-growing body of textual messages created by shifting populations of writers and interpreted by shifting populations of readers. . . . Quite literally, once a writer "posts" a text to a community, that text travels out into a digital world from which it cannot be recalled and within which it takes on a life of its own.[47]

If we define "text" broadly to include all verbal, aural, visual, and written exchanges, Burnett's formulation can help us understand more broadly how collective memory and identity "works" in portable communities. Indeed, the rhetorical construction of cyber-places (and all sociomental spaces) is a rich and fascinating area of inquiry; we need much more theory and research into the ways that words and accounts and stories inform the ways in which we create and approach cyberculture.[48]

Some blogs or websites are explicitly intended to serve as community memory projects, concerned with the preservation of cultures that may be marginalized or threatened in some way.[49] They may have as their sole purpose the creation and "housing" of collective memory.[50] Collective memory is highly fluid, extending backward and forward in time and stretching throughout space, with highly permeable boundaries and consisting of flexible, changing populations.[51] This has real implications for the definition of the life span—for individuals and groups. For example, blogs and websites are becoming heartbreakingly useful as places where people remember and memorialize loved ones after their deaths:

> I remember crying one night reading online that one of the founders of my listserv passed away. I witnessed and participated in my first electronic memorial service. People gave loving testimonial about how this dear woman had helped them in their time of need, how she touched their lives, how much they

would miss her and how glad they were that her suffering was over, in spite of the fact they had never met her. I will never forget the impact that memorial service had on me. As I read the messages on the screen I literally cried. . . . The list provides us with a very real outlet to express our condolences and a communal sharing of our grief. (SupportGroup1)

Just as technology permits the maintenance of relationships with people who are physically absent, it can facilitate the persistence of relationships even after physical death.

Audiences of all kinds comingle on publicly available memorial sites: mourners from various groups, interested visitors, the sites' creators. In a way, the deceased themselves become an audience. Photos of them may be in evidence. Messages are sometimes directed *to* them. We may freely express grief or guilt, share feelings, and even retell the circumstances of death; it becomes a kind of condolence book (such as those filled by mourners after the death of Princess Diana).[52] Some say things about or "to" an individual that would have been unlikely to have been shared while the person was alive. People who maintain these kinds of web memorial sites have reported an increase in angry public messages directed "to" the deceased—a phenomenon they call "dissing the dead."[53] We are now so accustomed to interacting via technology with others who are not in our physical spaces that it is probably only a small (and unconscious) leap to interact with someone who is *permanently* absent from physical space *altogether*. The deceased can remain very close to us on our mental maps; very much present in sociomental space.

There is theoretically no end to the length of time that a person can be so memorialized. Such memorializing extends, in effect, the story of a life. Now we can have a theoretically eternal sociomental existence:

A teenager in the town next to mine died in a car accident right after Thanksgiving devastating many people I knew. He had a site which he maintained before he died and after his death all of his friends left messages on his site telling stories and expressing their love for him and how much they missed him. It's five months later and people continue to leave messages—his sister maintains his space today. (MobileUser2)

We can see at such times the flexibility or malleability of the human life span itself. As profiles can be updated indefinitely and online identities can extend far into the future, the social life of the individual continues after death as people are remembered and memorialized indefinitely in cyberspace (and even before birth, as babies are anticipated and planned for in online sociomental spaces).[54] This extends the reach and the life span of the community in intriguing ways.

When many lives are simultaneously lost following traumatic events such as the Oklahoma City bombings or the September 11, 2001 terrorist attack, it is common for mass memorial sites to spring up. These give people a place to grasp and mourn the tragedy, to gather together as a community, to protest if they wish.[55] In helping us feel connected to others who feel the same sense of loss that we do, "the Web offers a malleable yet somewhat durable surface for collective commemoration over time, and some forms of historical reflection."[56] More scholarly work as to exactly how people use such spaces to pay tribute to others, especially en masse, to tell their stories and technologically "expand" their life spans is needed, but research thus far points to the increasing use of the internet for such purposes.

We are easily drawn into stories that have no definitive "end" and are propelled perpetually into the future. Discussion "threads" and community narratives can be continued almost endlessly, tapping into our human affinity for open-ended stories—narratives into which we can mentally insert ourselves long-term. As TVFan1 says of time spent on a favorite website, "Each of us tries to get the last word in, but you never can." One of my interviewees likened the format of these exchanges to a "soap opera":

> When [one of the members of our list] lost his wife, that hurt, because he had written so much about their struggle. It was almost like a soap opera with photos, etc. When she lost her battle, we felt we had lost a close friend. (SupportGroup5)

Seeing oneself as a participant in a long-term story can help a person feel part of something larger than oneself, something that matters and provides a sense of constancy and permanence. The telling and retelling of stories can ensure that the group will be there for a long time to come and extend far into the future—a definite appeal of the portable community. Theoretically, the community can go on forever—like an ongoing story with no fixed or scripted end, in the manner of a soap opera (or a blog, or a baseball game!). This lends authenticity to the community and enhances the realism of the experience, and allows us to feel that others are truly present to us, truly near.

PROXIMITY, PRESENCE, AND REALITY

In these sociomental spaces, where cognitive resonance, stories, and collective memories are created and preserved indefinitely, we can come to feel a real sense of the presence of and proximity to other people. The sensed presence of other people, or *social presence*, is the degree to which others in distant or mediated interactions, and our relationships with them, are seen as salient.[57] Put more succinctly, social presence can be thought of as the degree to which a person is perceived as "really there" in a technologically mediated setting.[58]

When people are not face-to-face, sociologist Christian Licoppe argues, "the [physically] absent party gains presence through the multiplication of mediated communication gestures on both sides, up to the point where copresent interactions and mediated distant exchanges seem woven into a single, seamless web."[59] As discussed earlier in the chapter, though people may be physically distant from one another, they can be very much cognitively present to one another. My interview subjects often spoke of absent others as being nearby, describing those with whom they are portably connected as being "there," being present, some "closer" to them than others. Like sociomental space (and social connectedness in general), social presence is a metaphor, one that is critical in helping us understand how people experience portable communities.

People can feel so close to one another, so strongly bonded, in portable communities because *proximity and presence are perceived by us in ways that transcend the physical*. When we connect with others, we experience real feelings of nearness to them: we may feel intimacy, love, happiness, anger. As we have seen, we sometimes even experience a special energy, a special chemistry I call *cognitive resonance* with certain of these others. In the mental map that guides our relationship with such people, they would be placed very close to us; we would feel warmth and heat, movement and life. These things feel real to us, much as they do in physical space, because our thoughts and feelings *are* real. As we see it, those whom we encounter in sociomental space are encountered *somewhere*. SocialNetworking4 puts it this way:

> [Chatting online] gives you the sense that we're just hanging out, hearing the day to day . . .

It feels like real life, because it *is* real life. Sensing the presence of others and feeling proximal to them is an important characteristic of the experience of using portable technologies.

Though we usually do not know exactly how many members of a given community are "out there" at any given time, a prospective member can generally gain a sense of the presence of "core" community members rather quickly. The core members are the most active participants in the group. They are the most likely to welcome new members or to monitor and enforce (formally or informally) the rules and norms of the group. Having had a stake in it the longest, they tend to take on the responsibility for safeguarding and communicating the collective memory and identity of the group.

But even those who lurk in the group or participate less actively help to shape it. Messages that circulate throughout the group are generally sent with the assumption that anyone and everyone in the group might receive and process them. In turn, these communications (whether they be textual messages, blog entries, pictures, or videos) are "read" or interpreted by many members of the community, even those who "lurk" and rarely send messages of their own. "The more an individual hears from others," asserts Caroline Haythornthwaite, a sociologist who studies social networking on the internet, "the more he or she may also hear about the network as a whole, and thus gain a sense of overall interaction."[60]

In the course of their communication—in writing or reading or speaking, and most critically, in *interpreting* any of these "texts"—the presence of the others in the community is discerned. For, Gary Burnett points out, "the transformation of distance into proximity" is "inherently part of the act of reading and is an inescapable element of all acts of interpretation."[61] As I have discussed, the cognitive activity involved in bonding without a visual draws us in; it involves us and helps us mentally envision one another. As community members perform the cognitive acts of reading, writing, perceiving, imagining, and interpreting, communities gain a "robust sense of themselves" and "an approximate but functional sense of their own populations."[62]

The community's membership thus becomes subtly but decidedly apparent. Spending time on her favorite website makes my interviewee AnimeLover feel like she is "not the only anime fan in the world, because there are so many people on this page." After starting an interactive website, she told me in some detail about the experience:

> I have my own homepage, and other people sign my guest book and write comments. I feel like I have an audience of friends, and I can stop by their pages and comment, or I can complain every day and they try to cheer me up. It makes me want to visit every day and always post something new . . . it's

also relieving to have that place to vent and be able to get feedback and sympathy.

As InstantMessaging4 relates:

> I do believe that IMing provides a sense of community for its users because we're all *at each other's fingertips.* (emphasis added)

Mobile phones, of course, are especially useful in helping people remain at one another's "fingertips"—and indeed, the sense of the proximity of people connected by mobile phones has been found to be particularly strong.[63]

Portable technologies help their users to remain in near-constant *ambient copresence.* This is an ongoing but "background" awareness of the presence or nearness of others.[64] Portable devices allow us to keep our channels to one another open nearly all the time if we wish, leaving "away messages" in text or IM form, checking in on these as often as we like. We also tend to use handheld technologies to stay in much more frequent contact with one another than we otherwise might. Some check in often with very short, frequent updates that simply convey that one is "around" or "there" (at this writing, one such application by which people communicate their most mundane daily moments to a community of others is called *Twitter,* and the act of doing so, *Twittering*). It is becoming common for groups of people (especially younger people) to stay in near-constant contact with one another this way, even as they are on the move themselves, in what has been described as a "full-on intimate community."[65] Many of my interviewees reported using portable technologies in this way:

> when I was at the height of instant messaging i was signed on at all times, and if i wasn't at the computer I would have had an away message up to tell them where i was, or i would just put a funny comment up. (SocialNetworking12)

> Sometimes there are moments when I am online and no one is around (they all have away messages up), but I'll still usually leave them a message and just say hello . . . (InstantMessaging4)

> Some people get a little carried away with that and tell their whole life stories on their away message. (Lurker3)

These methods serve to keep members of a community in cognitive proximity even when they are far apart physically, and to feel one another's presence almost continuously (for more on ambient copresence and constant availability, see chapter 6).

This sense of proximity—the presence and closeness of others—is a real and highly consequential phenomenon.[66] As SupportGroup1 told me, "This electronic list is a real life line." Our brains do not record "face-to-face" and "sociomental" experiences differently; whether an experience takes place in physical space or sociomental space, it is coded, processed, and stored in the same place in our brains (the hippocampus). Our brains, then, do not naturally distinguish the sociomental from the face-to-face; we must work to impose those distinctions on raw experience, and the process of doing so is imperfect. This is how movies can provoke feelings of real fear and horror in us (even causing us to jump out of our seats), novels can bring us to tears, and fiction in general can draw us firmly into the world that has been created. When our brains are engaged in social connectedness, whether face-to-face or sociomental, we treat it as though it matters to us, because physically, psychologically, and emotionally, it does. In the "realest" of ways, we are affected.[67]

The so-called reality of everyday life—the face-to-face "lifeworld"—is only one of the many realms or types of reality that human beings typically experience.[68] Many different types (or spheres) of reality have been identified as real, involving, and meaningful to us. Sociologists Alfred Schutz, Murray Davis, Peter Berger, and Thomas Luckmann, psychologist William James, and anthropologist John Caughey, among others, have all studied the "alternate" realities in which we routinely spend time and which are highly consequential for us. These realities operate as separate, distinct social worlds, each with their own norms, rules, and logics. They include such "worlds" as those experienced in dreams and fantasies, fiction, games, religious experience, erotic experience, drug-induced hallucination, and children's play. When we are absorbed in any of these, they can feel entirely, if temporarily, real.[69] Each carries an "accent" of reality, providing real (albeit different kinds of) textures to our lives.[70] Murray Davis reminds us that at any given time "we live not in one reality but in two (at least), and we continually alternate between them, often against our will."[71]

But we are not accustomed to thinking about reality and social life in this way. As a society we have come to define the "reality of everyday life," the face-to-face "lifeworld," as the premier, *real* reality. But, quite simply, there are many realities in which we circulate. Still, as Western society tends to privilege the literal over the figurative and the physical over the mental, a stigma remains attached to considering mental phenomena truly "real." Mental illness, for example, is often not considered "real" illness and is not covered in all health care plans. When we say that something is "all in your mind," we imply that something authentic is missing. And we commonly say that things that happen offline, face-to-face, happen "in real life" (it even has its own well-known online acronym, IRL). This way of thinking positions online life, and the life of the mind by default, as less than real. But mind and body, we must remember, are intricately connected. They affect one another continuously, as is indicated in psychosomatic illnesses or fatigue-induced mental confusion. Our minds and bodies "talk to" and inform one another all the time; they are a unit, finely meshed.[72]

Children, who often invent imaginary friends and toys that come to life, experience cognitive phenomena as real and are generally granted license to do so. But we are discouraged from such inventiveness as we grow older, to the point where our sociomental bonds become devalued as inauthentic, and we may feel silly or ashamed for making them, let alone considering them genuine. But just as children must learn to do, we must negotiate our own inner and outer worlds, our internal and external realities. Creative forms of self-expression and communication can afford ideal opportunities to reconcile the internal reality of the mind and the external reality of the physical world. Just as the security blanket and the "transitional object" do for the child, the portable community (and the relationships formed in it) may help us build a bridge between the world "out there" and the one in our own heads, because it is the province of both.[73]

In fact, many of us blur and play with the concepts of reality and fantasy rather freely. This doesn't necessarily mean that we can't tell the difference between reality and fantasy. It means that we have become so expert in forming sociomental relationships that we can be intrigued with them and give them meaning without necessarily confusing them with face-to-face relationships. In fact, we have become so good at this that we are able to see our favorite fictional characters as simultaneously real and constructed. Watching a movie does not confuse us as to the difference between the lifeworld and the fictional world; we can enjoy, even feel somewhat immersed in, both, simultaneously. Mentally approaching fictional characters as if they were real heightens the pleasure of that experience *and* gives us more practice in making sociomental relationships. We spend much time and energy making connections in portable communities. Now we need only give ourselves permission to consider the experience fully real.

The cognitive dynamics of portable communities are both real and powerful. They serve as no less than an infrastructure for portable communities. Indeed, I have argued at length elsewhere, a kind of cognitive infrastructure or scaffolding underpins all social structure and all of society itself.[74] The best way to examine this mental infrastructure is to expose it, talk about it, study it, make it somehow more visible.

In general, we are more comfortable talking about physical or face-to-face constructs as "real" than we are considering the sociomental "real."

But it is in understanding these cognitive dynamics and referring to them consistently as genuine that we destigmatize them. Even in the midst of writing this book, I have sometimes slipped and contrasted online phenomena with the face-to-face by referring to the latter as happening "in real life." (I know this because my students delight in pointing out this error whenever I make it.) But with sufficient understanding of mental phenomena, and enough practice using the vocabulary, we can remind even ourselves, when we forget, that *the mental is absolutely real!* Toward this end, I reiterate my suggestion that we avoid the use of the term *virtual*. I offer, as stated, *sociomental* as an alternative. Or since so much online and mobile activity is *portable*, that term might be another option.

It just isn't helpful to think of online and mobile activity as a species separate from, outside of, or less than, real life—not when real life (whatever that is) is drenched in cognitive activity. It is a false dichotomy. The mental *is* real, and it is all around us, not just in our heads. And the physical and the mental are inextricably linked. The cognitive dynamics that underlie human togetherness support and sustain all our social groupings; connectedness at a distance is merely the example par excellence of this.

Life in portable communities is cognitively rich indeed. In it, stories are told, thoughts are coordinated, and messages are exchanged and interpreted and given meaning. Real sociomental spaces are created and inhabited. People grow close to one another, cognitive resonance develops, stories and collective memory and community are built up. And in the process, the bits and pieces of a society are gradually, if invisibly, put into place.

3

FEELING CONNECTED
Emotionality and Intimacy

Communities have traditionally been sites of great emotionality and intimacy. In fact, the presence of emotion is critical to the building of any community.[1] Of the people I interviewed, 78 percent told me of intimate relationships or emotional experiences that had occurred in their online and mobile communities. Emotion can rise up quickly when strangers unexpectedly become friends, long-lost relatives are located, romances begin (and end) and move from online to offline and back again. Sometimes we make connections that would be unavailable to us face-to-face; other times we maintain and strengthen offline relationships.[2] But it is not unusual for an emotionally charged dynamic to develop when portable technologies are used; as WrestlingFan says, "I usually get excited when I am online."

Participation in portable communities, whether facilitated by computer, mobile phone, or other "smart" device, often gives people this kind of emotional "rush." AnimeLover describes it as such:

> One time I met a guy from Scotland online . . . we talked about anime and manga (Japanese comics translated to manga) and everything under the sun. It was crazy . . . it gave us a connection that we couldn't ever have had otherwise. I felt giddy like I was going on a date or something. It was surreal.

This "giddy," and indeed surreal, sensation can result anytime we take part in online and mobile activities—receiving a desired email, phone call, text message, or instant message; playing a game or completing a task; placing a bet, making a purchase, making a friend. The emotions evoked in these kinds of activities can be strong, intimate, and absolutely authentic, online as offline.[3]

When we think of a community, we think of people bound together not merely by similarity or circumstance but, usually, by some degree of emotional closeness or bonding (even if people are working together toward some utilitarian end). Sherry Turkle, a sociologist who studies science and technology, claims that we sometimes turn to technology because we *want* to feel something and believe that the technology will be a kind of conduit for feeling.[4] Love, hate, fear, rage—a multitude of emotions not only surface in online and mobile communities, but can strongly influence members, shape the identity of the group, and determine its degree of cohesion. In addition, trust and social support are both given and received, and even our moods can be deeply affected. In this chapter we examine emotionality and intimacy as it occurs in and shapes the portable community.

FRIENDSHIP AND INTIMACY

Many people do not hesitate to call those with whom they have become close in online and mobile communities their "friends." Friendship has always been established in the course of communication, through self-disclosure and increasingly intimate exchanges. When support, acceptance, loyalty, and trust are added to the mix, intimacy is firmly established.[5] And all of this can now be portably accessed and enjoyed.

Friendship, however, is defined and considered more broadly, more loosely, in portable communities. In electronic spaces there are all kinds of ways that people consider one another "friends." On many social network-ing sites, like MySpace and Facebook, people must declare one another "friends" to gain the most complete access to one another's pages. This practice has become so widespread that it has resulted in the recasting of "friend" as a verb (as in "he friended me so I friended him back") and a multiplication of meanings for "friendship." Social media researcher danah boyd has studied many social networking sites and outlined some of the reasons people "friend" others on these sites, including to gain access to see others' blog posts and profiles, to look cool or popular in others' eyes, to send a message to visitors as to one's identity and interests, and because it's easier to say "yes" when asked than "no."[6] We may also grant online "friend" status to offline friends, acquaintances, family members, and col-leagues. And it should be pointed out that not all face-to-face friendships are defined in the same way either.

Even outside of social networking sites, the term "friend" is commonly used to describe relationships that develop in online and mobile settings:

[On one message board] I pretty much know everyone. They are my good friends. Some I made online in one of the chat rooms and we became good friends. (InfoGathering2)

When I bounce ideas off someone, it's usually by phone or instant message . . . a nice immediate-feedback type of format that aids talking about how one of us is doing in general, which is really what makes him a friend and not just an online acquaintance. (Gamer1)

Some may even feel that they have become so close to others in their groups that they have become "family" to one another:

This listserv is my "cyberspace family," just as real as my own family . . . a huge electronic family, sharing the ups and downs of life. (SupportGroup1)

[On this discussion board] we are like a family. . . . We know about one another's personal lives and support each other on a daily basis. (ParentingForum2)

Though online and offline friendships are not qualitatively the same, and the term "friend" does not necessarily mean the same thing in both domains, there are certainly similarities. Behaviors and attitudes generally considered critical to maintaining the bonds of friendship—self-disclosure, supportiveness, positive social interaction—take place and are important both online and offline.[7]

When friendships develop online, they can be strong and supportive.[8] In fact, these bonds may be *more* engaging, less constrained by physical obstacles, than those friendships that develop face-to-face. Visual cues can in some cases encumber and distract from the essence of a relationship. Social attraction, involvement, and closeness can actually be enhanced when people do not have the means to see and touch one another.[9] Many people feel they can communicate more freely about themselves in the absence of physical distractions; one of the subjects in Andrea Baker's study of online romances noted that her partner may have received "a more accurate view" of her online than he would have obtained face-to-face.[10] Some feel they have more freedom, more choices online; others find it restrictive and more imposing than face-to-face interaction.[11] But for many, these portable kinds of friendships may be what sociological theorist Anthony Giddens calls "pure relationships"—bonds characterized less by traditional forms of commitment and more by the quest for personal satisfaction. A hallmark of the "late modern" age, pure relationships can be deeply intimate and "above all a matter of emotional communication."[12]

Online and mobile friendships and relationships exist in a wide range of forms—from "pure" and enduring to more ephemeral or situational, and from specialized (oriented toward getting to know one or two specific aspects of another person) to multiplex (oriented toward getting to know the other as a "whole person").[13] Even relatively fleeting social relationships can become quite intimate when those involved disclose much about themselves, feel that they are coming to understand one another, and feel good about themselves, the other, and the interaction. It is the positive progression of the depth of the relationship, more than its longevity, that renders it intimate. And such bonds can be highly durable, remaining meaningful and intimate for us long after regular communication has ceased (indeed, as we have seen, even after death).[14]

Still other online and mobile relationships are both intimate and long-lasting. The majority of close internet relationships remain intact for years. Psychologist Katelyn McKenna and her colleagues have found that we may like each other *more* when we initially meet over the internet than when we initially meet face-to-face. This may be because we feel that we are able to be more fully ourselves when we do not have to concern ourselves with being evaluated and judged on our physical attributes, and can instead feel more comfortable opening ourselves up to getting to know one another better.[15] "When we talk to someone in person," McKenna explains, "we pay attention to their subtle body language and facial cues that let us know how we are coming across. This fosters reticence in fully expressing our thoughts and feelings. The absence of these social cues in email can make people feel less inhibited about expressing what they are thinking and feeling."[16]

We can feel so close to those whom we have not seen face-to-face because we can learn quite a lot about others even when we only communicate textually.[17] When visual cues are limited or unavailable, we "fill in the blanks" ourselves, imputing physical and social traits to the other. We have become adept at approximating the expression of emotion online by creative keyboard use (inventing and using *emoticons* like ☺ or acronyms like LOL—laughing out loud—to convey emotion). And we are constantly gathering information about others from home pages, blogs, photos, webcams, writing style, mobile phone features, nicknames, etc.—there are innumerable clues to people's personalities online (perhaps more than are available to us in some physical settings).[18] As language and communication researcher Crispin Thurlow and his coauthors describe it:

Even with nothing but text, we can still tell a great deal about people from the language they use—their vocabulary, their grammar, their style. Besides, if we can't actually see social cues like age, sex and looks, we can always just ask. . . . This kind of direct request would seem pretty rude in FtF (face-to-face) communication but it's considered acceptable in CMC (computer-mediated communication).[19]

Using these strategies, some of the emotional content that might otherwise be technologically lost becomes "reinstated." Technologically mediated communication may seem cold, but we effectively "warm it up" in what psychologist Patricia Wallace calls the "socio-emotional thaw."[20]

As social creatures desiring intimacy, we will go to almost any lengths to create it, even in the absence of physicality. That's how smart and practical and imaginative we are. It is common practice to project onto our friends the qualities we want them to have or those that we may believe an ideal friend should have. We do this whether we are relating face-to-face or at a distance; interestingly, anonymity and the absence of visual cues can actually enhance the process of projection.[21] We want others to be like us in some way, so in the absence of disconfirming information, we often assume this similarity. Then, we feel close to these others—sometimes on this basis alone, almost as a kind of self-fulfilling prophecy. Assuming that others are like us and therefore would make good friends may seem like a rather undependable way to form a friendship. But even face-to-face, this is often how it's done.[22] And communication researcher Joseph Walther finds that technology-based relationships can be even *more* friendly, social, and intimate than those that develop face-to-face. He calls them "hyperpersonal."[23]

Technologies are continually being developed and popularized that approximate or reintroduce visual and sensory elements of the face-to-face experience to online or mobile connecting (camera and video phones, web cameras or *webcams*, streaming video, audio online). But, interestingly, these are not always the technologies that are most quickly embraced for the purpose of interpersonal connecting. Many of us prefer the greater anonymity of writing, especially for use in the early stages of relationships. Some shy away from using webcams in internet dating, psychologist Jeff Gavin has found, "because they feel it's important to not see their partners for some time—there is something special about text-based relationships."[24] Many people do not want portable communities to resemble the face-to-face variety too closely—they appreciate the difference. In certain circumstances, a bit of distance is preferred.

Some people feel less embarrassed, less self-conscious, when they can not be seen. Even in face-to-face copresence, they may avert their eyes when discussing something personal and emotional; they may not wish to be visually confrontational.[25] Being anonymous or physically invisible to one another initiates a "disinhibition effect," psychologist John Suler explains, for it provides "a built-in opportunity to keep one's eyes averted."[26] Even the temporary inability to see someone, as in darkness, can lower inhibitions. People meeting in a darkened room tend to disclose more personal information to one another and even to like one another more that those who meet initially in the "light of day."[27] My interview subjects confirmed this disinhibition effect, saying such things as this comment from MusicLover2: "I made some good friends online and sometimes it is easier to tell them your problems." This is akin to the "meeting on the train" phenomenon, in which we may feel more comfortable confiding secrets to a total stranger whom we do not expect to ever see again than to a loved one.[28]

> With some people it's easier to communicate online because you can be more removed from what you're saying, and that can be easier with strangers. Especially if you're trying to word something delicately. (WorkGroup1)

> By avoiding talking to people face-to-face anything people want can be communicated because they don't have to deal with the rejection, fear, and even harsh aspects of a face-to-face talk. (Blogger1)

We can be very much "ourselves"—we may even be *more* ourselves—when we are not entangled in face-to-face dynamics and pressures. "You may get to know more of the "real" person online than you would were you to initially meet him in person," McKenna notes, "because of the intimacy of online interaction."[29] Men, in general, seem especially appreciative of the opportunity to express intimacy textually,[30] though women evidence this as well. Research also indicates that women may be somewhat more comfortable, accomplished, and detailed in expressing emotions via technology.[31]

The absence of face-to-face interaction, then, often encourages the development of intimacy. In physical copresence, we frequently use appearance or social skills to judge others. We "filter out" those with whom we do not believe we would be compatible.[32] When visual cues are absent, we must pay more attention to one another's "insides" than their "outsides." Relying on shared interests and character qualities rather than physical (and perhaps more superficial) qualities, some technologically mediated relationships are happier and more durable than their face-to-face counterparts.[33]

Online relationships are most likely to survive when connectors meet at least occasionally face-to-face, which we are increasingly choosing to do.[34] Among my interviewees, 21 percent have met someone face-to-face whom they initially encountered online:

> When an opportunity arises I will try to have face-to-face meetings with the members of my listserv . . . (SupportGroup3)

> I have so many times talked with guys online and then at an event meet up with them and we become friends . . . (Social Networking12)

> Occasionally I make friends with new people on a site and sometimes meet up with them in public places like concerts or bars. It's good for networking. (Blogger2)

> [My message board] gets together in person a lot. Not only are there members throughout the country but also throughout the world. I know of people who travel all over to go to these gatherings. . . . We usually center these gatherings around our favorite bands, when they are playing in a bar or something. But some members of the board will open up their houses beforehand to have parties and hundreds of people will show up. (MusicLover4)

We can get to know people so well online that a sense of instant comfort can develop, easing the uncertainty of meeting face-to-face.[35] Describing the transition from an online to a face-to-face relationship, one of the subjects in Baker's study of online romance concluded that "the only thing that was different was that she had 'skin on.' " Another described the moment of physically meeting one another as "coming home."[36] Of course, it can be a bumpier transition, even a shock, especially when the relationship has been invested with strong hopes and expectations or has been idealized.[37] As E-dieter told me:

> My fiancé does not look like who I envisioned I would marry. If I had gone by an online description, I never would have started talking to him. As I got to know him face-to-face, his personality grabbed hold of me and then I found him incredibly attractive. We have many differences, but we learn from them. My world has expanded and now I have new interests thanks to him. And vice versa. My girlfriends set out online looking

for a guy that has the same exact interests as them—how is that exciting?

And SocialNetworking1 reports that

> I went on a horrible date with a woman I met [online] years ago. Typical web experience, her picture didn't match her actual appearance. Perhaps she thought the same of me. She was very "snarky" in person and not nearly as charming, so needless to say, no more web dates for me.

Many people who become friends or begin a romance online choose to meet in person as the relationship deepens, but as with all relationships, the outcome can never be guaranteed.

The online-offline migration works in the other direction as well. More and more, we choose to "meet" face-to-face companions and loved ones in online and mobile settings. In fact, we tend to go back and forth a lot—we may discover something newsworthy from a friend or family member in physical space, find out more about the topic online, relate additional information about it to someone face-to-face, return to learn more about it online, and so on.[38] Portable technologies make it convenient to stay in touch with friends and family and give us lots to talk about. This can help us maintain and strengthen our relationships. When we bring relationships with loved ones from literal space into sociomental space, we may find that social and emotional exchanges flow frequently and easily online:

> 99% of the people I interact with online are people who I am friends with in everyday life. I use it generally to find out what is going on in my friends' lives, to organize social gatherings . . . (MobileUser6)

> Texting is used for the most intimate relationships I have. . . . It's strange, actually, I only text people I'm very close to because we can rely on all our shared knowledge to fill in the gaps around the abbreviated message. (SocialNetworking1)

Portable communication technologies have become invaluable for sustaining face-to-face connections. In fact, the intimacy of friends who communicate both face-to-face and via mobile phones and text messages has been rated higher than those who communicate only face-to-face.[39]

However, it should be pointed out that it doesn't work this way for everyone. "If I desire a deeper sense of intimacy," MostlyEmail1 told me, "then online is no good." Not everyone is comfortable using on-line or mobile technologies to share intimacies. And at a certain point, to enjoy certain satisfactions, there can be no substitute for the physical touch. Face-to-face togetherness can be preferred for a variety of reasons, as WorkGroup1 points out:

> For more sensitive conversations (either negative or positive) I strongly prefer talking in person, or on the phone, because tone of voice can be easily misunderstood in emails.

For SocialNetworking2, the lack of visual cues diminishes the emotionality of the exchange:

> You lose the emotion and non-verbal communication that comes with face-to-face conversations or oral communication, because you lose the ability to hear pitches in the voice, the gestures, etc.

And some of us simply use portable technologies strictly for instrumental purposes (for more on practical uses of portable technologies, see chapter 5). In these cases, emotionality may be reduced. Still, research indicates that the use of portable technologies generally becomes more emotional and expressive over time as people grow more comfortable with such technologies and use them to meet their deepest needs.[40] These include the giving and receiving of trust and support.

TRUST AND SOCIAL SUPPORT

Contrary to what some believe, online and mobile communities can be extremely trusting and supportive environments. Of course, as with intimacy, this is not always the case—not everyone feels comfortable trusting at a distance. But living in a modern technological world actually requires us to place our trust in all kinds of people whom we don't know or don't know well: legal and medical professionals, the producers and distributors of products and materials we use, members of financial, cultural, govern-ment, and military institutions, and so on. We must, to some extent, place our trust in others whom we will never meet, people who may be part of large, far-flung social systems. Unless we are to constantly question our reliance on all of these people, to constantly question our safety and social stability, we must assume that regulation and regulating agencies—bureaucratic rationality in general and all kinds of experts in particular—will do their jobs more or less properly. Social interaction, and indeed society itself, "would be imperiled were individuals unwilling to trust the legions of physically absent others on whom they are dependent."[41]

So we do. As a matter of course, we tend to invest trust in absent others. It is not a great leap to consider that we might then come to form bonds of trust and even deep commitment with those whom we come to know personally, albeit at a distance, in portable communities. As we get to know these others, sometimes even contacting them directly, we try to assess their reliability. We attempt to deduce (sometimes by even explicitly researching) something of their reputation.[42] We may decide to use some of the widely available online services that gather and present opinions, comments, or rankings to help us do so. Though we usually do not know the individuals or companies that perform these rankings, they are often very much trusted as well (Google's ability to rank results is generally considered accurate and trusted by users). Our confidence and sense of being supported in online and mobile spaces is therefore increased, our fears of exploitation are decreased, and, whether well placed or misplaced, we tend to trust in absent others.[43]

Furthermore, anonymity in and of itself acts "as a foundation for the building of trust and establishing real world relationships."[44] The tendency to disclose personal information anonymously online can lead to familiarity and reciprocal self-disclosure. All this can actually happen much more rapidly online than off. Trust can develop surprisingly easily and quickly in the absence of face-to-face interaction.[45] InstantMessaging2 writes of how technology assisted her in disclosing her sexual orientation to her parents by creating a trusting space in which they could more easily have the difficult conversation that served as a precondition for the development of trust:

> After many months of secrets and lies, my parents finally confronted me about my sexuality in an email. I am a lesbian and had been in an abusive relationship. I had just not been able to find the words to talk with them. Finally, I got an email saying, "hey, we know you are gay and it is okay," basically. I responded immediately from my own apartment with tears streaming down my cheeks, but I was so relieved it was finally out in the open. Eventually I would have told my parents, but it would not have happened so quickly and candidly.

The support that this young woman received had been eluding her in her physical encounters with her parents. But once established, trust can engender ongoing social support.

Social support flows freely and frequently along the pathways of on-line and mobile communities. This is especially true when the community consists of a flourishing network of strong, "bonding" social ties.[46] Support can be formal or informal, emotional or instrumental, and as freely given as received—if not more so.[47] As SupportGroup5 says, "I can give and take blessings . . . or good karma." Reciprocity, even purely altruistic behavior, is very much in evidence in cyberspace. Those who have been helped in the past may wish to help others. Social status and social capital are also gained from helping others. We may hope that doing good work or being knowledgeable may bolster our standing in a community. Again, this is as true online as offline.[48]

Support groups accessed by online and mobile technologies can provide many of the benefits traditional self-help and mutual aid groups provide: problem solving, information sharing, catharsis, the expression of feelings, and mutual support and empathy.[49] Members of these kinds of communities tend to share both confidences and fears, often with candor, warmth, and humor. Communities of support enable participants "to pool their 'collective intelligence' about many things," says Patricia Radin.[50] In her research on online medical and breast cancer communities, Radin describes the significance of such communities in the lives of people who have "few other outlets to speak honestly about their feelings and emotions, to get access to the experience of many others facing similar predicaments, and to learn about a wealth of resources" and survival strategies.[51] SupportGroup1 discusses what it is like to need social support urgently and have nowhere else to turn:

> This listserv has been a lifeline to me, especially when I was first diagnosed and could not rely on the few minutes I had with my doctors to help educate me on what my disease was and what I might expect. . . . Until you learn the new language, it is disconcerting to be facing a disease and not know how to communicate effectively. It is like being dropped into a new country and told your life will depend on how quickly you learn to communicate in this new language—it is a life altering experience.

Radin notes that the importance of even one person helping another person at a highly sensitive or difficult time should never be underestimated. And this happens time and time again in portable support communities.[52]

My interview subjects speak to the importance and emotionality of exchanging social support in trying times:

> Going online or texting makes me feel better if I'm sad and want to talk to someone or need support. (BookLover1)

> It's relieving to have this place to vent and be able to get feedback and sympathy. (AnimeLover)

There are also emotional benefits to helping others:

> It allows me to feel better about myself, in part just because helping others always feels good, and in part because helping others is the best way to ease grief. (SupportGroup3)

> It has made me realize how blessed I truly am. There are so many people in worse situations than I am in. (SupportGroup2)

> There is a feeling of camaraderie and belonging [with online groups], a real sense of being able to help so many more people than ever possible before the advent of the internet. (SupportGroup8)

ParentingForum1 reports that she has felt the need to decrease her participation on her electronic mailing list except for "when I can respond to questions I can help with."

Providing support to others can be emotionally satisfying and can help individuals feel that they, in turn, have been helped:

> I get to know how my words have helped others, and that is a profound privilege to know that. It brings meaning into my life, and support to me when I need it. . . . I feel better when I'm told that one of my posts has helped someone else. (SupportGroup6)

Disinhibition often results in people sharing very personal stories with groups, even very large groups, of others. Many people report that they join online support communities for just this reason; a degree of anonymity is appealing to them. On the other hand, others view such groups with suspicion; they worry about deception, misinformation, identity theft, and other potential hazards.[53] But for many, the need for and convenience of the group tips the balance. They report feeling part of an ongoing cycle of giving and receiving support. This taps into the very human desire to help others and, when the time comes, to be helped.

To be able to access a support group anytime, anywhere, is especially appealing:

> I have a lot of people in my family who are suffering from cancer, and have joined chat rooms dealing with cancer. Talking to people is an emotional uphold. If I am having a bad day, I go online rather than call up someone which is very odd. It has become something I do on a daily basis. It doesn't matter what time it is someone is always online, and I don't have to feel like I am disturbing someone by calling. (Chatrooms1)

In addition, family members and support professionals benefit as well, often using portable technologies to offer what social work professor Joan Beder calls "cybersolace" in the form of generalized social support, practical information, shared experience, helper therapy, empowerment, professional support and advocacy efforts, and support groups dedicated to the caregiver. "It is a technology built on emotion and caring," Beder says.[54]

Participating in online support groups, especially with great frequency, can be so emotionally involving as to be draining. SupportGroup1 notes the importance of taking occasional breaks:

> The sometimes detailed descriptions people offer of their treatments, fears, pains and other stressful and painful aspects can be depressing to read and for those who are already carrying a heavy burden, emotionally or physically, it can be real tough. The listserv is not for the "faint of heart." I have had to take a break. You need to be fairly strong emotionally and physically to stay with the list day after day.

SupportGroup3 tells me that

> I recently have greatly reduced my use of the listserv since it is just too hard to watch cyberfriends suffer.

To be sure, the high levels of intimacy and emotionality to be found in portable support communities can be overwhelming. ParentingForum1 relates how "very upset and discouraged" she was when online friends of hers passed away. But participation in such groups can also help alleviate stress and improve physical health, particularly when people have some experience using technology and feel good about their technological capabilities.[55]

Portable connectedness often results in needed, deeply felt support. We are both producers and consumers of this support. Trust is easily, often freely, exchanged in portable communities. In the process, our moods and behaviors can become very much affected.

HAPPINESS AND HOSTILITY: THE MODERATION OF MOODS AND BEHAVIOR

Portable technologies are often used to help us moderate our moods, emotions, and behaviors. It is a general expectation that time spent in online and mobile device use will improve our well-being in some way; indeed, users' moods tend to improve following time spent online. We tend to become less tense, less angry, and happier when we have spent time in portable communities.[56] My interview subjects had much to say on this topic:

> Checking my email and getting a funny joke or kind words can always lift my mood. (OnlineStudent4)

> Text messaging and IMing can really affect my mood, the same as calling someone on the telephone . . . (MovieBuff)

> Text messaging boosts my mood sometimes—if I'm tired or frustrated at work and a friend sends me a text message about going out somewhere I get excited and the day seems to go by somewhat quicker. (MobileUser2)

> Some days it boosts my mood. Specifically when I check my email and get messages from old friends or job offers or discount offers or newsletters. (WrestlingFan)

Just as we have friends whom we look to in physical space when we need to be cheered up, our portably accessed friends and acquaintances can help us moderate and improve our moods as well.

The exchange of intimacies and social support alone can be real mood boosters. And simply knowing that we are on someone else's mind can make us feel good . . .

> It is uplifting getting comments or messages from friends who you would normally not get the chance to ever communicate with due to everyone's different lives and busy schedules. . . . It makes you feel like, hey, this person was thinking of me and

you feel liked by others and that always puts a smile on my face. (TVFan2)

. . . as can learning something interesting and new:

> Text messaging usually puts me in a better mood because in a way it's like receiving a letter from someone. It's always exciting to read something new from an important person in your life. (MobileUser4)

. . . and having the opportunity to share our feelings or "vent":

> the parenting group i belong to can make me feel better. they are fun and with some of them it is like being with IRL (in real life) friends. i can vent if i had a bad day or share something good. (ParentingForum1)

Sometimes it's as simple as it is for SocialNetworking1:

> Having a snappy comment or a perfectly worded post makes me feel good!

Any application of portable technology used in these ways can help people enhance their moods. Computer-mediated communication, and in particular the sharing of written disclosures with an interested audience (as on blogs or discussion boards), has even been demonstrated to alleviate depression.[57]
Of course, our moods can also be negatively affected, because social interaction is always layered with both positive and negative dynamics:

> I don't know of anyone who during the course of a rough day at work doesn't appreciate a good joke or funny memory sent via text message. Thinking about it though, I would assume that it can be used in a negative aspect as well, though I have not experienced this. (MobileUser6)

Experiencing sadness online or feeling upset is certainly not unknown:

> Emails and texts with friends/family often make me feel better [but] occasionally emails will also make me feel worse, when someone communicates something sad, or feeling sad, or feeling upset with me in an email. (WorkGroup1)

. . . nor is self-consciousness. . . .

> Honestly, sometimes looking at others' picture albums on their facebook pages makes me feel worse because some of my friends have really great pictures up of all the places they have traveled and famous people they have met and I don't have any of those. It can also make me self-conscious about my weight and looks when I view their pictures but this also happens when I look at magazines anyway. (Lurker1)

. . . and sometimes we simply miss having face-to-face contact:

> Sometimes talking online or text messaging is depressing because there are moments when a human voice needs to be heard instead of a beep to indicate you have something to read. (MobileUser2)

As we have seen, the exchange of social support can sometimes be an emotional burden:

> Sometimes, the list does get to me and I do feel burdened and "get down." . . . it is very hard to read of so much pain and suffering . . . (SupportGroup1)

Though online and mobile groups can bring us social support, they also expose us to the difficulties, hardships, and frailties that exist in all communities and forms of human relationship.

Sometimes, a negative mood can be the result of the limitations of online and mobile technologies:

> It makes me angry when I have an online class and can not get in touch with a professor. (Chatrooms1)

Message boards can become very negative in tone and disturbing to members:

> Just as in all human communication, whether it may be face-to-face or electronic, conflicts do arise. . . . I sometimes read posts that become volatile and antagonistic if the members don't agree on something. (MusicLover1)

> If some "newbie" posts that they don't like a particular song or they think a certain album stinks or such, people will post

about how that person doesn't know what they're talking about. And then the name calling will start. (TVFan1)

Some people are very harsh and will put you down without a care. (MusicLover3)

It is, of course, impossible to control completely the substance or outcome of interpersonal interaction (for more on the control of social interactions, see chapter 7).

InstantMessaging2 wonders whether we sometimes take advantage of the lack of face-to-face accountability to behave inappropriately:

On occasion I have found myself in a fight or argument with a friend via text-messaging or IM'ing. Maybe because it is "easier" to say everything this way rather than face-to-face. Maybe it is used as an avoidance mechanism.

Others agree:

I definitely save confrontations for the computer. (Mostly-Email4)

My e-boyfriend and I would typically argue online and we were more hurtful to each other over the internet because we weren't looking right at the other person or hearing the other person's reaction. (InstantMessaging5)

Online I can get on my friends' nerves and they just log off without saying anything and I can not hold it against them because I really don't know for sure why they logged off. Whereas if I'm talking to them on the phone I can tell if they hang up on me, or face-to-face if they are not paying attention. The feeling of being online is sometimes awesome, sometimes depressing. (InfoGathering2)

As we have seen, darkness and distance can give rein to less inhibited behaviors. Moods are not only moderated but can be manipulated in portable communities.

Much research has been undertaken on conflict in online communities.[58] *Flaming*, or the posting of hostile remarks, is fairly common in online social interaction. Believing that community members will never meet face-to-face seems to encourage some to express negative thoughts and feelings, often for no apparent reason.[59] *Trolling*—posting derogatory or irrelevant remarks to bait community members into pointless discussions—also occurs online fairly frequently. New members to groups (*newbies*) are often advised to take special care to learn the norms of the group before "jumping in" haphazardly to avoid the possibility of being flamed.[60] Participants actually work quite hard to keep their groups cohesive in the face of a "flame war" or the activity of a "troll." They adopt strategies of ritualizing or normalizing the harsh or negative remarks through withdrawal, offering apologies, denouncing the offender, posting poems, attempting mediation, showing solidarity, or joking.[61] Some treat the "flamer" or "troll" with a patronizing or classist tone. Handling group conflicts is a process in which all members have a stake. If unchecked, the result can be nothing short of the dissolution of the group.

In an attempt to maintain community solidarity and to keep flaming and trolling to a minimum, many, though by no means all, message boards are moderated. Key tasks of moderators are to prevent and handle flames, to deal with trolls, and to keep the group on task. Though a skillful moderator can help the group stay focused, several of my interview subjects expressed disappointment with boards that are "overmoderated." When dissent is too easily squelched, many feel that one of the prime advantages of the online community—the ability of members to interact freely and get to know one another—is compromised:

> Some of the moderators on the board take their "job" far too seriously. Someone could write a post comparing an episode of a show to something that is going on elsewhere in the world, and a moderator will flag it and shut it down because it is talking about politics or religious subjects. I mean c'mon people!! It's just a message board, why do people have to take things so seriously? (TVFan1)

Disruptions, confrontations, flaming, and trolling have become expected and generally accepted, if still rather emotionally affecting, aspects of the cyber-community experience.

Online and mobile connecting can bring us up or bring us down. Though our moods are more often improved in portable communities, they are of course sometimes made worse. We can be harassed or find the experience disruptive or harmful. We can spend so much time on the computer, phone, or PDA that we find the overall effect mood-depressing (for more on the downside of technology use, see chapter 4 on technological dependency and chapter 9 on social problems and inequalities). It was even pointed out to me by Gamer1 that studies like mine may be part of the problem!

The feeling of being stuck to the computer is pretty depressing. I'm not sure why it is, but sometimes I just feel like I can't pull myself back into reality. I end up answering messages from researchers about internet life . . .

Our moods can certainly depend on the specific outcome of an online or mobile activity:

I go on a site where I can play poker against other people throughout the world. And depending upon whether I win or lose, my mood can be affected. (TVFan1)

And as SupportGroup8 reports, we can become physically affected as well:

My butt gets sore sitting here so much, which makes my lower back sore, and it keeps me from doing other things (reading books, magazines, chores, paying bills, etc.)

But there are also those who claim that "going online does not affect my mood in any way," as MostlyEmail2, told me. Or, in the words of InstantMessaging7:

Going online doesn't make me feel better or worse, really. The only instance where being online will make me feel better is if I go online after being off for an extended period of time (more than a week).

There is still much we have to learn about the ebb and flow of moods while online, and the ways in which the act of spending many hours with a computer or portable device can affect us emotionally, psychologically, even physically.

Because online and mobile communities are sites in which people come together in real human fellowship, emotions are constantly displayed and exchanged and heightened. Because our access to others is ongoing, we may begin to expect others can help us moderate our moods at any given moment. We may even come to depend on technology to help us access others *in order to* have a particular feeling at a particular time. This can have far-reaching implications, Turkle warns:

For some people, things move from "I have a feeling, I want to call a friend" to "I want to feel something, I need to make a call." . . . When technology brings us to a place where we're used to sharing our thoughts and feelings instantaneously, it can lead to a new dependence, sometimes to the extent that we need others in order to feel our feelings in the first place.[62]

Portability in communication technology inspires more frequent social interaction than has been possible since the onset of the industrial age. With industrialization, we are often geographically distant from one another and must travel great distances to be at one another's sides. Portable technologies, in a very real sense, bring us closer to one another—always mentally, though sometimes physically as well. It is tempting to want to "close up" that distance and become even closer by reaching out to technologically accessed others to help us express, echo, validate, even *feel*, our feelings. Continued research is needed into the effects of exchanging feelings so profusely and instantaneously, and of the cultivation and management of so much emotion via technological facilitation.

Emotions serve as a kind of "glue" for a community, especially an online or mobile community, which can not count on the small moments of face-to-face experience—the subtle gestures, tones of voice, body language, touches—to cement it. It is as though the cognitive infrastructure of a technologically gathered grouping becomes surrounded with "emotional glue." With the presence of emotion, trust, and support, the group gains heft and solidity; it becomes a true community. Along the way, we can find our moods moderated and manipulated.

This exchange of emotion can easily result in the "rush" of social connectedness spoken of so frequently by my interviewees. To know that social bonds can be readily formed and more or less constantly accessed, and that intimacy and trust and social support is available nearly anywhere, anytime, is powerful knowledge indeed and can be quite exciting to the would-be social connector. Even participating in my research gave one of my interview subjects a bit of a "rush":

> It makes me excited, because I get to talk about [my online life] in a legitimate study. (AnimeLover)

How much more, then, must time spent in a long-standing portable community, readily accessed by portable technology, give participants an emotional charge exactly when they may need it most. As Social Networking8 says, "Sometimes when I get back to my room I just move the mouse and go to my favorite site and check my profile . . . and it's like someone has left me gold or something!"

4

PLAYING AROUND

Fun, Games, and Hanging Out

Sometimes we use portable technologies to come together simply, seemingly uncomplicatedly, for the fun of it. About a third of American internet users, at this writing, go online on any given day for no particular reason—just to pass the time or have fun—and two-thirds have done so at some point.[1] Indeed, 67 percent of my interviewees described time spent in portable communities as occasionally or often "fun." Using portable technologies tends to be enjoyable; we often expect to have a good time when we engage with these technologies.[2] Whether playing games on a computer or mobile phone, gossiping, joking or flirting via email or text message, or just "hanging out," perhaps on a social networking site, time spent in portable communities often has a light, playful dynamic.

Play is activity that is bounded apart from everyday life, separated from pressures and obligations. It is freely chosen, noninstrumental, often absorbing and escapist.[3] We can play games that have rules and structure (as in basketball or Monopoly), we can play informally (as in tossing a Frisbee or dashing in and out of the ocean's waves), we can play music (on a musical instrument or on the radio). We may inject the spirit of play into some other act (by kidding around with our friends or delivering a sarcastic remark that might be considered an insult but for its playful delivery). Or one may simply be in a playful state of mind. All of this is play. Play is a common and consistent aspect of life across all cultures.[4]

Play is fundamentally relational; it "is a 'mode of being' subsuming both the object of play and the consciousness of players."[5] As we play and have fun, we hone skills, interact with others, and form relationships. We model and experiment with social behaviors. Play is how the child determines its self and its place in the world, and it does much the same for those of us that are older.[6] Playing is "just as important as reasoning and making" to our humanity, says historian and pioneer of game studies Johan Huizinga.[7] In other words, play can be serious business.

There is a "primacy of play" in human-computer interaction.[8] Sociologist and researcher of communication and cyberplay Brenda Danet notes that computer-mediated communication is an inherently playful medium in which

> not only hackers, computer "addicts," adolescents and children, but even ostensibly "serious" adults are learning to play in new ways. . . . Digital writing is inherently playful, first of all, because the medium, the computer, invites participants to "fiddle," and to invoke the frame of "make-believe."[9] When this frame is operating, participants understand and accept the meta-message "this is play."[10]

According to Danet, this playfulness is fostered by four interrelated features: ephemerality, speed, interactivity, and freedom from the constraints imposed by physical materials. Her analysis might be extended to include all kinds of constraints that are imposed by physicality in general. As portability enhances each of these features, it can provide more opportunities for, and enhance the impact of, play.[11]

Gaming, joking, gossiping, flirting, and hanging out are all activities designed to be playful. But in the process of having fun we can cross a line and cause harm to ourselves and others. As offline, "too much" is not necessarily better. A habitual user of portable technologies can find that excessive use brings about dependency, overstimulation, fatigue, depression, or other negative outcomes.[12] In this chapter we examine both the lighter and more serious sides of fun, play, and just plain "hanging out" in portable communities.

GAMES

Game playing is a highly popular activity in online and mobile communities. A game is a formal system of play that is rule-based, behavior-constraining, and voluntary, with a quantifiable outcome. It is artificial and is bounded "outside" ordinary life, although, obviously, it occurs in the real world, and its outcome can very much matter in a very real way.[13] Games provide challenge in a structured setting, neatly satisfying our often contradictory needs for novelty and for familiarity. They can be exciting and highly absorbing in nature, or have a calming effect. When portably accessed they can provide nearly unlimited ways for those of us who want to pass the hours in play to do so, regardless of where we are or the presence or absence of someone with whom to play.

In addition to the inherent enjoyment of the game, we play games online and on mobile devices to feel connected to others, for games can be highly social. They provide a platform upon which to make new friends and solidify existing friendships. Some people prefer to play online and mobile games with others physically by their sides, either participating in the game or watching the player play.[14] Danet notes that one can play a game online "even when there is no apparent partner to play with. One is playing with the program and the machine . . ."[15] The technologically generated atmosphere of the game, populated as it may be with characters, plots, and activities (even if crudely depicted), can temporarily "stand in" for face-to-face friends, situations, and environments and fulfill at least part of our human desire to be with others.

Both men and women enjoy playing online and mobile games. Such games are proving increasingly welcoming to women, though female characters are often sexualized and underrepresented relative to males.[16] Relative to console-based video games, games that can be accessed by online or mobile devices tend to be more flexible in their game options and less violent in nature, with female characters that still may be sexualized but less stereotypically so than in other kinds of computer games.[17] Some women prefer text-based gaming environments because they tend to be more character-driven rather than action-driven, and find long-term gaming enjoyable. Some women like playing games or simply spending time in settings in which their gender can be disguised.[18] For men and women alike, video, computer, and online games can provide a reliable source of fun, recreation, and social interaction, accessible most any time of the day or night:

> I haven't sat down and played a tabletop roleplaying game for years, mostly because of time issues. But on a [computer] bulletin board [game], people can game from around the world, without much worry over timezones . . . (Gamer1)

Physically dispersed people can use portable technologies to enjoy games together in comfortable and comforting environments.

Gaming brings people into the player's life, people who may turn out to be helpful in all kinds of ways:[19]

> When I am playing a new game and I need a hint to advance in the game, I check the boards regularly. If I need a hint or there's a game glitch, someone has usually run into it before me. . . . Although my gaming is essentially a solitary pursuit, I

can rely on the hard-core players and contributors to the bulletin boards to provide on-the-spot help, which, in turn, helps me enjoy my game more. (Social Networking1)

Being able to access help—and *other people* to do the helping—makes the portable gaming experience that much more satisfying. As discussed in chapter 2, we tend to find it interesting to meet people with whom we know we have at least one thing in common; it provides a starting point for future social interaction. A great deal of "underground" networking surrounds many technology-based games: people meet up, exchange strategies, divide the game "labor," and get to know one another further in online, mobile, and face-to-face venues.[20]

Game play can be the basis for the building of large player communities. Many millions of people worldwide take part in *massive multiplayer online role-playing games* (also called MMOs or MMORPGs). In these games, individuals interact to construct immersive games, story lines, and social worlds that are highly intricate (much as they do in MUDs—*multiuser domains*—and Second Life, in which participants also create complex social worlds that they then inhabit online). These games and online worlds are structured so as to be intensely involving for hours at a time, and, often, for years and years.[21] Perhaps most strikingly, though, they are sociable activities and environments. According to Anthony Fung, a communication researcher specializing in the study of these kinds of games,

> online game players must learn to join communities, make friends, deal with other players, and cultivate a socially acceptable virtual character because there are informal tribal rules in the game, as well as strategic methods for characters to "raise their levels" within the game. . . . This cultivation of community is not only instrumental online but is also beneficial to players in daily life and in further extending their social lives.[22]

It is those participants who make friends along the way that are most likely to remain active in online games.[23] "The main reason people are playing is because there are other people out there," says social media and gaming scholar Dmitri Williams. "People know your name, they share your interests, they miss you when you're gone."[24]

Being part of these social worlds and being able to find fellow players at any time of the day or night can provide several satisfactions:

> you are never bored . . . really you can use your mind, alleviate boredom, hang out, and make life more interesting. (Gamer2)

When I asked how gaming can accomplish all of that, Gamer2 replied:

> It is just fun. I love playing games and I enjoy the fact that yahoo gives their games to us for free. . . . When playing games online you usually can always win especially if you try hard. So this alleviates boredom and it makes me happy when I win.

People tend to associate positive feelings with online gaming, more often labeling such experiences "pleasant," "exciting," and "challenging" than "frustrating," "boring," or "stressful," even though conflict and frustration abound in most games.[25] When frustration does occur, it may actually motivate players to improve their game play so as to try to escape that state of mind.[26]

Games often provide a series of "interesting choices." These choices may inspire us to try to master the game. The mastery of these strategies—and of course, winning the game—is usually exceedingly pleasurable. Games can make us happy, even when there is little or no skill or effort needed to win, as in a game of chance.[27] This may be because games tap into the brain's natural reward circuitry providing psychological and emotional satisfaction.[28]

And the more challenging aspects of game playing are, in their own way, benefits. There are difficult tasks that must be mastered, strategies that must be crafted, decisions to be made, and whole social environments to be created. All of this helps players develop and practice skills that will be valuable in other settings, proposes science and technology writer Steven Johnson:

> In every second of a video game you must evaluate its state and make quick and accurate decisions. So you are getting smarter in terms of the ability to soak up a lot of data on the fly, then figure out the best strategy and decide the best path forward. . . . To be able to sit in front of a new piece of technology and manipulate it without reading the manual is an important form of intelligence, one that we need more and more to be economically competitive. Games are doing a great job of training kids to do that.[29]

Online or video game playing can be a useful skill for those looking to successfully compete in the modern workplace and society. It can teach decision making, problem solving, discipline, and delayed gratification—as good game players learn that making adjustments to repeated failures can be the road to success.[30] It can even teach the scientific method, for

[t]o succeed, a player must establish a hypothesis about some aspect of the game, test it, and evaluate the results of the experiment . . . a video game often hides its rules, revealing them only as the player figures out how to unlock the game's secrets. And when that happens, a game player can experience an ecstatic Archimedes moment.[31]

According to global business and communication researchers John Beck and Mitchell Wade, video game players are more likely than nongamers to develop critical skills and traits—cognitive, creative, competitive, *and* social—that will aid them in the workplace.[32]

Video and computer gaming can inculcate players with values like competitiveness and obedience. There may be unintended consequences, though. Game players generally win by submitting to the rules of the game, not subverting them. They may devise fairly creative strategies to win the game, but these strategies exist within the framework of the game; gamers are not permitted to rewrite the rules of the game itself. "Follow orders, and you'll be fine," says journalist Chris Suellentrop, summarizing what he calls the "winner's ideology." Game playing may strengthen players mentally, he argues, but it dampens their ability to innovate, to be flexible. "Don't worry that video games are teaching us to be killers," he concludes. "Worry instead that they're teaching us to salute."[33]

Games can teach a wide variety of values, or none at all; they affect people very differently, especially as they range so widely in their content, structure, and in the level of involvement they inspire. But when especially challenging and involving, games can induce the highly desirable mental state that psychologist Mihaly Csikszentmihalyi calls *flow*—a state in which we become so utterly engaged in an activity that we become immersed in it, our sense of self-consciousness disappears, and time falls away.[34] Flow is most often experienced when a task is voluntary and is undertaken at the level of difficulty appropriate to one's skills. As Johnson says,

the brain is most likely to be engaged in learning when we are working within the "regime of confidence": not too difficult and not too easy, but on the edge. Games are specifically designed to work exactly on that line.[35]

Many games, especially those that we play with someone else at the same time (synchronously), induce flow. In these games, action and awareness become fused. It becomes easy to lose all sense of time, suddenly discovering that hours have passed. An engrossing feeling of freedom, efficaciousness, and happiness that draws us ever more deeply into the social world of the game (or the community surrounding the game), can result.[36] However, games do not always inspire this kind of experience.[37]

There is much we have to learn about the complex experience of game playing, particularly as the playing of online and mobile games now infiltrates so very many physical spaces (restaurants, buses, even classrooms). Computer game researcher Espen Aarseth calls for an interdisciplinary approach to digital or "postdigital" game studies that would examine the intersections of the design, art, and actual playing of online and mobile games. Such findings could translate to cyberculture studies in general, he argues, for "there is probably not one characteristic commonly ascribed to cyberculture that could not be found in the gaming sector."[38] There is much to learn about playful activity in portable communities including, but not limited to, games. For we do not always desire the intensity or conflict of a game. Sometimes, we want an experience that is more quietly sociable, more lightly playful. These more laid-back kinds of activities are also found in online and mobile communities.

JUST "HANGING OUT"

We spend a great deal of unstructured time in portable communities just "hanging out," with no agenda in mind other than, perhaps, to fill the hours.[39] We may be procrastinating, putting off doing something we don't want to do, or we may want to feel a little less bored or lonely. Teens often use instant and text messaging because it enlivens their lives; increasingly, younger and older adults do as well, also using social networking sites, blogs, podcasts, and other applications just to give us themselves to do.[40] As SocialNetworking7 says, "If I am bored, I am probably online."

But online and mobile communities are also spaces for relaxed socialization, for hanging out. Sociologist Ray Oldenburg calls areas like these "third spaces"—spaces that complement the home and the workplace by providing a "third" place in which individuals can come and go as they please in relaxed, casual fashion. In online and mobile "hangouts"—as in their face-to-face counterparts: pubs, coffee shops, beauty shops, and the like—people can have fun and socialize with minimal obligations or social entanglements.[41] And because they are portable, there is always someone "there."

We need places to hang out—places other than home and work in which to spend time and unwind. They provide a low-key environment where we can relax and meet different kinds of people. Some gaming sites function as third space hangouts. Communication and gaming researchers Constance Steinkuehler and Dmitiri Williams note that

Because virtual worlds are perpetually accessible (barring the occasional software update) and played in real time, participants are free to log on and off as they see fit. Populations are commonly heaviest in the evening, reflecting the free time of most youth and adults, but there is always someone on at any hour.[42]

The game site or online world (such as Second Life) can thus be seen as an ideal hangout, where "one may go alone at almost any time of the day or evening with assurance that acquaintances will be there."[43]

Many find this relaxing:

I find that the internet is a place where I can unwind and not think of work because I am at freedom to go to the sites of interest to me and that have no connection to work whatsoever. I find myself chatting a lot with old friends and shopping. . . . I am so busy throughout the day that when I get home all I want to do is relax by being online. (WrestlingFan)

I think the computer and internet is to my generation (or at least me) what the TV was to my dad's generation. It's what we do to relax. (TVFan1)

And as mobile phones permit us to catch up informally with friends nearly anywhere we (and they) are, they in effect turn any place into a "hangout":[44]

Of course it's fun—it's a way that allows me to talk to my friends all the time, what more could a girl ask for? And if I can't talk to them online I know I still have my cell phone to say "hi." ☺ (InstantMessaging4)

When I spent a lot of time in the car, I liked being able to make calls on my drive home—like IMing, it allowed me to catch up with family and friends during a time that would already be wasted. (OnlineStudent1)

We can also hang out when and where it might otherwise be impossible to do so . . .

Sometimes if I am waiting for something and have time to "waste" I will call someone I was supposed to be in touch with to use the time. (ParentingForum1)

> I log on to my blackberry and email friends and family while I am waiting for an appointment or if I am in an airport waiting for a flight. (Activist1)

> I found myself so busy and far from my friends that late night conversations via cell phones was the only method to keep in touch and catch up. (SocialNetworking2)

. . . and "steal" time from other more productive uses:

> Sometimes when I need to get away from work for a while I will take out my cell phone and go down one by one through each of my saved numbers. (MobileUser3)

Stealing away via portable technology allows us to carve the time and space of a hangout *out of* our work or home time and space. This subverts the usual spatial separation of hangouts in physical space. Now, our sociomental hangouts can intersect with physical ones, and be all the more easily and frequently accessed. We can integrate these spaces or separate them, partitioning our time among these spaces or multitasking, giving portions of our attention simultaneously to, say, a face-to-face interaction, an email discussion, and a mobile IM session. Technology consultant Linda Stone calls this giving "continuous partial attention" to more than one activity; a special kind of multitasking that is especially common among youth.[45]

Sometimes, we use our portable technologies specifically to alleviate boredom. Of my interview subjects, 51 percent reported using them for this reason:

> I go on [my favorite social networking site] when I'm bored. . . . If I'm extremely bored and it's after 9:01 p.m. (when minutes are unlimited) I sometimes browse the games available on my phones but only choose the ones which are free. (MobileUser2)

> I go online when I have nothing to do. Sometimes I am in my room and am bored out of my mind so I go on the internet just to check my mail (which I hardly receive) and to see who is signed on. (InfoGathering3)

> I frequently will connect to the internet to check my email or make phone calls to family and friends to alleviate boredom. (WorkGroup2)

> When I had free time and was "available" I used to check out
> the profiles and photos of single men on dating sites just for
> something to do. (MobileUser1)

As MusicLover2 sums it up:

> I participate in these groups when I am bored, just to look for
> something to do, also I made good friends . . . and it is fun.

It is possible that we are less and less tolerant of old-fashioned boredom
now that we have the means to alleviate it (or to try to do so) any time
of the day or night:

> For the most part I think involvement in these groups
> has helped to curb my boredom with everyday life. Not to say
> that my life is boring, it isn't, but it makes life just a tad more
> fun and exciting to feel that I am part of something other than
> just family and work. (TVFan2)

And this is exactly what third space hangouts do for their denizens. Regardless of how satisfying home and work lives may be, most of us want
our lives to consist of something more. Participating in face-to-face leisure
activities (sports, clubs) can be just as demanding as work or domestic
responsibilities. Modern lives are increasingly filled with things to do. By
definition, hanging out requires little more of the individual than to simply
contribute one's presence. Comfortable, inviting environments, constantly
populated with friends, where little is required or expected—portable communities serve as excellent hangouts.

Hanging out can take the form of *lurking*—visiting a community on
a regular basis but not posting or contributing very often, if ever:[46]

> On the message boards my posts are very random and not often
> so I don't really interact with anyone. But I do get a lot of
> pleasure from frequenting these online groups, it is entertain-
> ing to me to read the thoughts and ideas that people have.
> (TVFan2)

> Mainly, I like reading other people's opinions of songs and the
> members of the band. (MusicLover4)

> I am online for days at a time, but I am often idle. (Social-
> Networking2)

Because it is not obligatory to converse actively with others in most portable communities, people can lurk indefinitely. In fact, it is not uncommon for the vast majority of members of a portable community to be lurkers. We lurk for many reasons, including pleasure, the desire for privacy or safety, the belief that we do not have the time or the need to become more involved, and personality factors such as shyness or a preference for emotional detachment.[47] Nonetheless, many lurkers consider themselves full participants in their communities and feel that they know other members well, even considering them friends.[48] As MobileUser2 describes it:

> I don't directly interact with people on the site but I send messages and somewhat lurk . . . [and] it gives you a sense of connectedness.

Silently and invisibly, many lurkers simply enjoy the feeling of being connected, of being a part of a community.

Hanging out is good for us and even contributes to the health of a society.[49] It helps us feel that we belong to something larger, something communal, something fun. We enjoy spending time in these spaces and finding friends there. And just as in our brick-and-mortar hangouts, conversation is an important part of the experience. Though online talk can certainly be serious, it is often light and playful as well—humorous, gossipy, and/or flirtatious—when we are hanging out and having fun in our portable communities.

PLAYFUL TALK: HUMOR, GOSSIP, AND FLIRTING

When we're playing games, having fun, or just hanging out in our portable communities, we like to talk—even though this "talk" often takes written form. Communication scholar Nancy Baym, who researches many aspects of computer-mediated communication, points out that writing things to one another via technology is "in many ways more like conversation than writing because it is interactive, relatively spontaneous and generally unplanned."[50] Conversation easily and often turns playful—humorous, gossipy, sometimes quite flirtatious—in portable communities. This may be a reflection of boredom or of having extra time on one's hands, but it also speaks to something deeper and more substantial. For even seemingly inconsequential interactions reflect and shape our deep desire for social bonding. These "less conspicuous forms of relationship and kinds of social interaction," sociological theorist Georg Simmel asserts, "produce society as we know it . . . [and] incessantly tie men together."[51] As we chat, laugh, gossip, and flirt, what may masquerade as lighthearted exchange can actually serve to tie us tightly together.

We exchange small talk to relax, enjoy ourselves, bond with one another, and relieve tensions. Online as offline, gossip, humor, and flirting act as social lubricants. They help us begin and end conversations, soften requests, and indicate that we can address the other's point of view. They may pave the way for further intimacy. Thinking similarly about what is funny, strange, or ridiculous, or sharing private in-jokes, can signal like-mindedness, a similar philosophy of life. This is perhaps why we tend to be attracted to those with senses of humor similar to ours. It helps us establish and enjoy the common perspective and common ground so important to social bonding.[52]

Humor and playful conversation can enliven the often mundane nature of everyday life; they give us something interesting and slightly out-of-the-ordinary to focus on. Light humorous banter may be "informationally inconsequential but socially important," Gary Burnett and Harry Buerkle claim.[53] Lighthearted exchanges aid in self-presentation, facilitate the creation of common understandings and cognitive resonance, and help generate group solidarity.[54] They can also be used as a way of "conveying serious information without appearing to do so," says sociologist Michael Mulkay.[55] In her study of humor in an online discussion group, Nancy Baym notes that

> [h]umor more often than not occurs in explicit reactions to others' messages, and humor often invokes past group discussions. Humor is thus one of the ways in which participants blend the group's discourse into a unified whole. What could be an ongoing stream of messages with little coherence are transformed into group history and interpersonal contacts. This does not happen in all Usenet groups, and may be one reason other groups do not develop the sense of community found in the community studied. . . . The humor is a joint production, which the audience not only understands but helps create. This joint authorship enhances group identity and solidarity.[56]

Humor creates a friendly, sociable atmosphere and makes the group seem fun and appealing—all the more needed in the absence of face-to-face visual and verbal cues.

Gossip is also common in portable communities. According to social anthropologist Kate Fox, this is because it serves many critical functions for us:

> Gossip is the human equivalent of "social grooming" among
> primates, which has been shown to stimulate production of
> endorphins, relieving stress and boosting the immune system.
> Two-thirds of all human conversation is gossip, because this
> "vocal grooming" is essential to our social, psychological and
> physical well-being.[57]

Gossip—and other light forms of "grooming talk"—is a playful and enjoyable way to keep up on the latest news and gather interesting bits of information. It can also elevate one's status by being in possession of, or sharing, information that not everyone could easily obtain.[58] Incidentally, men seem to gossip as much as women, especially when utilizing technology. In an echo of general gender conversational patterns, women's gossip tends to be more detailed and to take place with same-sex friends and family, while men tend to gossip more with colleagues, partners, and female friends.[59]

Portable technologies lend themselves to the sharing of gossipy conversation. Mobile phones are often used to quickly share or "preview" interesting bits of information for later face-to-face discussion. Text messages may announce topics that will be expanded on in greater depth during future encounters (Fox calls these "trailers"). Along with IMs and short emails, mobile text messages are well suited to communicate these kinds of short announcements and playful pieces of business.[60] They make it easier for us to maintain frequent contact with a wide social network, "even when we do not have the time, energy, inclination or budget for calls or visits," Fox points out. "Texting re-creates the brief, frequent, spontaneous 'connections' with members of our social network that characterised the small communities of pre-industrial times."[61] In a preindustrial age we were in much more frequent contact with most of our family, friends, and acquaintances. With portability in communication technology, we can be in frequent contact again.

Gossip can spread very rapidly via portable technologies. Social networking sites can be hotbeds of informal, chatty exchange. According to Blogger1,

> [My social networking site] is basically just for people to be nosy
> and find out what others have been doing or are doing with
> their lives. . . . Mainly girls use things like this to be nosy and
> talk bad about other girls. It is like an online gossip community
> and most of the time the girls can't get enough of it.

InstantMessaging3 notes that

I was going through a break up and my ex and I have a lot of the same mutual friends. My ex just happened to see one of our friends online and chose to talk to him about what was going on. Later when I signed on, this same friend sent me the whole conversation . . .

The public, portable nature of these technologies means that keeping things private has become more difficult to do (for more on this, see chapter 6). In addition, we can never be sure whether our messages have been electronically passed along to someone else.

Like gossip and humor, plenty of flirting takes place in portable communities.[62] Flirting "along the friend/lover boundary" flourishes in portable communities due to the safety often perceived in distance and anonymity, the relaxed, informal nature of hanging out online, and, in no small part, the ambiguity and "deniability" that the use of text can provide.[63] As flirting is not an explicit sexual activity but a possible and uncertain prelude to it or substitute for it, it is actually quite a complex act. It is both stimulating and playful and contains ambiguous sexual/romantic overtones, yet with lower stakes, at least initially, than in full-fledged affairs. Some flirt in the hope that a relationship or affair will develop, some to discover whether others find them attractive, others simply to pass the time of day, while for others it may be an either deliberate or automatistic form of playfulness. With flirting so much in evidence, relationships that are not physically sexual can easily acquire sexual overtones that range from mild to pronounced. Due in part to the ambiguity and prevalence of all this flirtatious talk, friendships "tinged with sexual nuance" are very much in evidence in the modern age, online and offline.[64]

We learn to flirt online much as we do face-to-face, by adopting a "playful and ironic" self-description, attitude, and tone, and, if writing, attempting to translate those things to text.[65] As one's tone of voice is absent in the written word, we must spend extra time when conversing textually to ensure that our words match our intent as closely as possible. In addition, such matters as the pace of the communication and pattern of the turn-taking take on additional significance. As OnlineInstructor told me, "a delay in responding can be misinterpreted as a lack of interest or as pondering on the last message"—both dubious messages to send in the course of flirting. Additionally, if one has not responded to another person's emails, posts, text messages, phone calls, or other overtures after what one considers to be a reasonable time, the online relationship or flirtation may be considered ended.[66]

As with gossip, the secrecy of flirting can be fun, exciting, and of real consequence. According to sociologist Christena Nippert-Eng, when secrets are shared and kept, social bonds are created, for the people involved are brought together in a special type of relationship. People are connected in the sharing of private, secretive information, in engaging in secret activities, or when certain aspects of the interaction are kept secret. Others are excluded when a secret is kept; a boundary around those "in the know" is drawn, often very firmly so. Every relationship consists of some secrecy, since it is impossible for us to share every thought, every action, even with those loved ones that are closest to us. The pattern of secrecy in a relationship, then, tells much about the relationship. In what we choose to share and what we choose to withhold, secrets define and circumscribe our relationships.[67]

Humor, gossip, flirting, sharing secrets, and other means of conversing promote and bring about social bonding, on and offline. They provide a way for people to interact, become informed, feel important, and feel that they are in community together. Plus, they are often described as lots of fun:

> It is definitely fun and exciting each time I log on! (Wrestling-Fan)

> I IM my friends and we chat online just for fun. (Mobile-User1)

> It is fun to share how my daughter is doing and to share everyone's lives. (ParentingForum1)

> [My anime community] is just fun for me and gives me a rush I can't explain. (AnimeLover)

The "rush" so often found in online and mobile connectedness and discussed in chapter 3 can now be more fully understood. It is a combination of cognitive resonance, emotionality, intimacy, fun, and the sociability that can be found in gaming, hanging out, chatting, and conversing in a portable community. Engaging in enjoyable, sometimes exhilarating, activities with others who find them to be similarly so tends to be enjoyable and exhilarating in turn. Instances of experiencing this sense of excitement can be multiplied many times over as we join numerous portable communities. But, of course, there can be serious consequences as well. Fun and play can be pursued excessively and inappropriately. Fun can actually have a seductive, rather dark, allure, which should certainly be considered alongside the lighter side of life in portable communities.

THE SEDUCTIVE ALLURE OF FUN

Having fun is crucial to a well-balanced, well-lived life. But activities like gaming, flirting, hanging out, and playing are not always benign. While they are often beneficial—and surely provide the "rush" AnimeLover and many others have spoken of—they can have a harmful flip side. For one thing, it is easy to lose track of how long we have spent in playful activities. We can become immersed in what online game and virtual reality researcher Nick Yee, who has surveyed more than 40,000 MMO players in his Daedalus Project, calls "problematic usage"—the overuse or abuse of technology.[68] We can also use technology in ways that are harmful to others and to ourselves. All interactional life is a balancing act among tasks and activities, among work and play and rest. When should we be concerned that we are not simply having fun in our portable communities but are engaging in activities in a troubling way or to a troubling extent? Put another way, how much is too much?[69]

Ironically, it is when we are at our most relaxed and playful that we may actually be becoming most deeply involved in portable communities and the social worlds created. There has been much debate as to whether and in what circumstances technology use can be said to be "addictive." Experts disagree as to whether this is the best way to describe extreme amounts of use. But it is becoming clear that certain online activities such as game playing, gambling, and pornography can stimulate us psychologically with the potential to draw us in deeply over the long term.[70]

There are face-to-face analogies here. Offline as well, excessive time spent in such activities as game playing, gambling, pornography use, alcohol or drug use, shopping, even sex—in any of a number of behaviors that can lend themselves to compulsivity—causes concern. These behaviors can be encouraged or facilitated in any number of settings—face-to-face, online, mobile. But the accessibility and relative affordability of portable technologies has permitted increasing numbers of people to become more easily enmeshed in such activities, including teens and adolescents. As writer Julie Hanus says regarding the accessing of pornography via portable technology:

> In this Internet-ready age, technology tracks a highway to our living rooms, bedrooms, and offices. Cell phones, iPods, PDAs: There seems to be no limit to the places porn can appear. . . . Pornography is no longer simply out there for those who seek it; it is *in here*, inside computers, inside telephones, inside homes.[71] (emphasis added)

In addition to pornography, technology permits drugs, gambling, and consumerism to more easily enter our homes and personal spaces. With all of this product so readily available, some of us overindulge, some of us try to ignore it, some of us become desensitized—but all of us are affected.

There is much controversy as to whether internet use itself belongs on a list of addictive or compulsive behaviors. That which is often labeled internet addiction, abuse, or dependence may more accurately be considered symptomatic of other problems such as depression, sexual disorders, or loneliness.[72] Some experts, like Dmitri Williams, wonder whether the addiction paradigm is appropriate. "If a person was reading novels excessively, we'd be less likely to call that 'addiction' because we value reading as culture," says Williams. "We see game play"—and, I might add, other so-called nonproductive activities—"as frivolous due to our Protestant work ethic." Continued research would surely help us understand and delineate this important issue more completely, for, as Williams reminds us, "it's not the role of science to guess or bet."[73]

To be sure, excessive internet use has been associated with various problems of an educational, physical, psychological, and interpersonal nature, all of which are worthy of additional study. Experts in addiction tell us that a behavior becomes problematic when the individual becomes enmeshed in a downward behavioral spiral characterized by mounting life problems, the failure of coping skills, and intensified cravings.[74] According to Nick Yee, problems in everyday life are the best predictor of excessive use in game-playing settings. A person who plays games in order to avoid or escape other problems, rather than simply for entertainment or socialization, may be more likely to become excessively involved in gaming. This does not mean that people with problems necessarily turn to excessive online game playing, but that many who are troublingly engrossed in games may be escaping other problems. Some people "feel they lack control," Yee explains. The game, then (and perhaps other online involvements) gives them "social status and value" that can be especially enticing to those with serious personal problems.[75]

It should be noted, though, that more people in general report positive than negative effects of internet use overall, especially when used for social purposes.[76] The effects of internet and mobile use on the individual depend on the nature of the activity being undertaken, the outcome of that activity, and the motivation for and extent of use. Put another way, the many varied uses and applications of portable technologies will always interact with individuals' needs and motivations to produce different experiences and effects. We should not expect or look for outcomes that are negative; this would be as gross an oversimplification of the phenomenon as it would be to assume that face-to-face interaction is necessarily negative simply because troubles and risks abound.[77]

Some prefer to call excessive internet use a "dependency." In the words of psychologist and professor of gambling studies Mark Griffiths, use of the internet can become "a non-chemical dependency that may be passive (ex: surfing) or active (ex: chatting)" and in its extreme form, can feature the core components of dependency: salience, mood modification, tolerance, withdrawal, conflict, and relapse.[78] Those who are "super-connected" can certainly develop a feeling of dependence on the technology and, in a deeper sense, on the human connectedness it facilitates at any time. Psychologist Robert Bornstein calls this a "dual dependency"—the need to have portable devices nearby all the time *and* the need to have other people always be reachable.[79]

In Yee's study of MMO gamers, 45 percent describe themselves as "addicted." Robert Wood and Robert Williams's large-scale international study of gambling online found that 42.7 percent of internet gamblers can be classified as "problem gamblers." Of my interviewees, 28 percent used the word "addiction," unprovoked by me, to describe online and mobile technology use, by themselves or someone else, they had witnessed:[80]

> I'm addicted—I "need" to be online and I don't really know why. . . . it is hard sometimes not being around a computer because you are wondering what your friends are doing or wondering if someone you haven't spoken to in a while has IMed you, so the negative is how badly I "need" to be online or near a computer. (InstantMessaging4)

> The groups are quite addictive. You find yourself posting more often and looking to see if someone has sent you a message. You can easily spend an hour on the site without even knowing it. (BookLover2)

And SocialNetworking8 confides that

> computers I am realizing have slowly taken over my and my friends' lives. . . .

Some do not use the word "addiction" specifically, but are clearly concerned about the amount of time that they or others spend online:

> I feel that you can get caught up with the whole online life. It does take a lot of time and time just flies by. (Message-Boards2)

> I could spend a week straight online if I thought I would survive that way. (Anime Lover)

With all the information that is available online, you could stay
forever there. (MusicLover2)

Without doubt, there is an allure to spending time in these resonant,
emotional, intimate, playful spaces.

Others worry more specifically about the effects of time spent non-
productively than about addiction per se:

[the parenting message board] is a time waster. it takes away
from the time I should be working at night. all the moms there
feel the same and we complain about the time we waste talking
together. (ParentingBoard1)

I have a tendency to occasionally check the board while at work
so this can hinder my productivity. (MusicLover1)

It has made me feel worse, especially when I looked at myself
and I have spent all day online, without showering and without
eating just snacking. I go, what a waste of my time, I have
missed a beautiful day. (MusicLover2)

. . . and in the course of worrying, the word "addiction" resurfaces:

While I did engage in internet dating it was very time consuming
and sort of addictive. . . . I checked it all the time. My house
cleaning suffered! (Social Networking4)

My participation in online groups has given me the opportunity
to stay in touch with others and have some fun. . . . However,
it can be addicting . . . my work suffers sometimes because I do
not focus my full attention on it. (Lurker1)

Some feel that rather than turning too often to more playful pursuits,
they (or others) may be becoming excessively productive (i.e., "addicted
to work"), given the opportunity provided by portable technologies for
one to work almost anywhere, anytime:

Since I got a wireless laptop, now I am on more and more and I
have to tell myself to put the computer away. . . . I had to build
a home office just to work in because I was working all over the
house, anytime, anywhere, with the wireless . . . (Activist)

It is easy to feel that "technology is to blame" for things that happen in technological spaces. This is called *technological determinism*, and is frequently invoked when people feel surrounded by or unduly influenced by technologies that they feel they can not escape. Some feel they must make endless decisions with regard to technologies that they do not fully understand, and that these decisions have grave consequence. Although technologies are tools invented by people for their own use, they can seem to overwhelm us and "take us over."

But it is important to note that for most of us, after a bit of a "honeymoon period" in which portable technologies are overused or misused, such technologies tend to be successfully integrated and "folded into" our everyday lives:[81]

> When I first started engaging with the online community it was different because I spent more time online than anything else . . . I balance the rest of my life now very well. (Chatrooms1)

> I used to keep myself signed online all day long so that when I got bored I could talk to someone. Now I only sign on when I am bored, and usually get bored again quickly and sign off. (WorkGroup4)

> I used to spend a lot of my free time online, probably 3-4 hours a day, sometimes more back when I was in high school. . . . Now I am on only sporadically, most of the time when I am bored at work . . ." (WorkGroup5)

Sometimes we simply get tired of a technology or application:

> My friends from high school and I had blogs for about a year and a half. There were probably 7 or 8 of us and I read them all a few times a week. I don't think we have posted in 3 or 4 months and I only check about once a month now for new posts. (SocialNetworking7)

> Last year some of my friends had blogs and I read them sometimes—but most of those fizzled out b/c it was too time-consuming and really just served as another thing to add to our procrastination. (Lurker3)

> I think it was around 4 or 5 years ago when I realized that I was spending too much time roaming around the web. Specially

I was too frequently visiting the news sites only to realize I'm getting the redundant information on each site. I believe this changed my online behavior. Now I try to spend as little time as possible on any website. When I log on to the web, I usually have an intention . . . I don't use the web to kill my time anymore. (InfoGathering4)

Chatrooms2 sums it up:

These experiences are interesting to me in the beginning, but lose their appeal after a while.

We may use portable technologies frequently, casually, and in nearly all places and social settings, but we also generally try to integrate them with activities that have a more explicitly physical basis in a manner that is more or less workable for us. And for most of us it seems to add more to our lives than it subtracts, as we shall explore further in the next chapter.

Most of my interviewees worry less about compulsive, pathological, or "addictive" behavior in portable communities than about their propensity for old-fashioned procrastination—using ever-present portable technologies to avoid doing things that they otherwise might (or should) be doing:

I can watch TV and talk on the computer for hours at a time. It is an easy way to procrastinate, and take my mind off of all the things I have to get done. (InstantMessaging1)

I tend to make myself more bored by doing such a thing as connecting to the computer to hang out, even when there is no one to talk to me. I think I could use my time wisely in other areas if I didn't have a computer always available to me. (MobileUser4)

Homework, it seems, is quite often a casualty:

the only thing that suffers from being online is my homework, but I know that I would find something else that would keep me from my homework if I wasn't online. I'm pretty good at using every possible excuse to keep from doing homework. ☺ (SocialNetworking5)

I often go online or call people when I'm bored just as something to do. Usually around homework time, the computer can be a procrastination aid . . . (BookLover1)

As a professor of sociology, I must admit that this one's my favorite:

> The only thing that may suffer a bit is homework time, but real obligations don't suffer! (MovieBuff)

But, of course, if a person wants to procrastinate or put off obligations, there will always be a ready excuse. And the convenience of portable technologies may result in a net savings in time overall. I can't imagine how long it would have taken to write this book on a manual typewriter—like childbirth, I seem to have conveniently forgotten the difficulty involved in such an activity. As WorkGroup1 says, "in the long run it all saves me time and energy."

The impact of spending considerable time in portable communities extends well beyond the immediate fears and concerns of individuals. A culture is constantly in change, reflecting the habits and preferences of its members. Portable technologies have resulted in the mainstreaming of certain things once thought of as more "deviant" because the greater availability of these things online and in our homes has resulted in their greater availability in the surrounding culture. Gambling, for example, is increasingly tolerated and even promoted by parents as a positive social activity for children. Consumerism is rampant in technological societies among young and old.[82] And a kind of "culture of pornography" that extrends well beyond pornography seems to be infiltrating a number of social spheres:

> We gobble up illicit video clips of the latest scandals; cooking shows linger lazily on luscious ingredients before making the baking appear to be seamless; video games nod to a classic porn format: little bit of plot followed by action-action-action. There's lifestyle porn, disaster porn, food porn, and more. At times, we scarcely notice. In place of XXX marquees, we see utterly predictable provocative billboards. Standing in the grocery store checkout line, we skim right over *Cosmopolitan*'s explicit cover lines.[83]

What is considered pornographic symbolizes a society's attitude toward sexuality, and in a larger sense, what it decides is acceptable to expose, to reveal, to fetishize, to consume.

The ubiquity of explicit materials "loitering in everyday portals"—like the casual approach to gambling, drugs, gaming, consumerism—is reflective of a "distinctly modern shift in attitude," Hanus argues. She assesses the scope and impact of the technological delivery of pornography into private spaces by noting that "when technology began delivering pornography into our homes, it also secured pornographic patterns of consumption into our technologically mediated lives, blurring the boundaries between porn and what becomes porn simply because of how we expose, experience, and consume it."[84] We now see pornographic patterns of consumption of all kinds of technologically delivered content and we see voracious (if not necessarily addictive) appetites for all kinds of technologically delivered experiences. It should not surprise us, then, that portable communities can be deeply harmful, wildly playful—and just about everything in between (they can even, perhaps, be benignly neutral). There are endless untold effects to be examined. Chapter 9 takes a look at some of these, but much more research is needed into the consequences of the production and consumption of all this technological culture.

Because it can be so seductive, easy, and so much fun to go online or connect with one another via technology, we (and our children) may indeed become immersed, even engulfed, in behaviors like gambling, pornography use, compulsive sex, compulsive shopping and consumerism, the pursuit of unhealthy relationships, alcohol and drug use, and other behaviors. All of these things have been problematic before the advent of portable technology, of course; "blaming the technology" is not a solution to any social problem. It would be far better to seek to understand the forces and desires that underlie technology use. Although there does tend to be a honeymoon period in which online and mobile use is especially seductive, we can not assume that harm will necessarily follow the introduction of a technology to one's everyday life. Yet we are also well served to recognize that for some, a healthy balance is not reached: for some, a few drops of gambling or shopping or porn quickly becomes a flood. Online and mobile activities can indeed have long-term effects that we are just beginning to understand, especially as they are becoming more normative and we may be becoming more desensitized to (even jaded by) them. These are among the topics most deserving of further study.

Some of my interview subjects confided their concern that we (and our children) have come to expect and even to require instant stimulation at all times—to never be bored. To be sure, the generation of younger people that have grown up with these technologies at their disposal are so accustomed to them that they may be less tolerant than older people of boredom, slow pacing, of things that are not so much fun. They have likely been exposed to so much gambling, pornographic or sexualized or violent images, drugs and alcohol, and especially consumerism, that they can not possibly process it all. They hang out on their social networking sites, chat spaces, and mobile phones hour upon hour, interacting with all kinds of people (though more often than not these people are offline friends as well) and witnessing all kinds of behaviors. It remains to be seen how all of this will be folded into their lives as they create and live in the portable and face-to-face communities of the future, and research into this should be undertaken as well.

But, with all of this playful, seductive, even dangerous activity occurring in portable communities, we should keep in mind that people regularly and reliably use them to be productive, to form social relationships, and to make plans to gather together face-to-face. Portable technologies help us make and maintain social relationships and social culture in the ways that we (individually and collectively) choose to use them, in the ways that work for us. They provide us with unlimited social spaces in which to play games, hang out, chat, and have fun. Cyberspace and portable communities are enjoyable places to be. But in the end, most of us prefer to be together in physical spaces for work and for play, and to form communities and relationships—seductive and otherwise.

5

SOCIAL NETWORKING
Convenience, Practicality, and Sociability

To be a member of a portable community is first and foremost to be a social creature—one who actively forms connections and relationships with others. As Blogger3 told me, "When I think of being online I think of connectivity, which to me means staying in touch with those I love." Portable technologies provide us with the means to form *social networks* with others practically everywhere we go, all the time. Yet another spatial metaphor, social networks are in effect mental pathways along which social support, resources, and, I would argue, cognitive connectedness and emotional experience, can "flow." Social relationships, too, seem to flow along these pathways, much as cable television signals flow along cable lines. Using these networks, old friends can be located, new friends and acquaintances and lovers can be found, information and products can be obtained, and tasks can be accomplished. And we can stay better connected with others, including those whom we also see face-to-face.[1]

Social networks cross-cut all our communities. We may think in terms of groups, but we live in networks, sociologists and social network experts Barry Wellman and Keith Hampton argue. In their words:

> Communities are clearly networks, and not neatly organized into little neighborhood boxes. People usually have more friends outside their neighborhood than within it: Indeed, many people have more ties outside their metropolitan area than within it. Their communities consist of far-flung kinship, workplace, interest group, and neighborhood ties concatenating to form a network that provides aid, support, social control, and links to other milieus.[2]

Social networks provide communities with structure, and individuals with opportunities to form ties and connections with others. The concept of the network can be extremely useful in helping us understand the dynamic of interpersonal sociability that we see in such abundance in portable communities.

Those who have access to technologies of portability possess the tools that enable these networks to be formed at any time, in any place. As WorkGroup1 tells me:

> These technologies have made it possible for me to stay more connected with more people than I could otherwise, and to meet people who can help me personally, academically, and professionally.

Or, as WorkGroup6 puts it:

> I like knowing I can be in touch with anyone, anywhere, at any time.

Time and time again, my interviewees spoke of the enhanced sociability provided by technologies of portability such as computers, mobile phones and devices, and PDAs. In this chapter we look at the social dynamics associated with networking in portable communities: the pleasures of discovering old friends and new lovers (or new friends and old lovers), coordinating and "microcoordinating" activities with relative ease, and accomplishing a wide variety of tasks. We will examine sociability in online and mobile communities, the use of social networks for convenience, for dating, romance and sex, and for learning, working, and getting things done.

SOCIABILITY

A prime use of social networks is socializing—making new friends, reconnecting with old ones, and spending time with family both physically near and far away. Though email, IM, mobile phones, and webcams are all useful in this regard, social networking sites that have the express purpose of helping people find and associate with one another specialize and excel at this. These sites create spaces for us to share our thoughts and all kinds of information, express emotions, have fun, and become, if we wish, intimately involved with one another. In these spaces, we feel that we are meeting up *somewhere* (hence the widespread use of such spatial metaphors—most obviously "MySpace") in sociomental space, and we get to know one another in all kinds of ways.

Looking up old friends and faraway family members is an extremely popular use of online and mobile technology:

> [Social networking sites] help you stay connected to people, and are practical in the sense that you can find people from your past. (WorkGroup4)

> I believe [social networking sites] do serve to network people together. I have visited them many times to reconnect with old friends . . . People tend to drift apart as their lives change. I like to keep in touch with many old and new friends as a way of living a more enriched life. (WorkGroup6)

The pleasure of reuniting online with old friends—sometimes unexpectedly and serendipitously—was mentioned by several of the people I interviewed:

> Using one site I was able to reunite with a friend from kindergarten. As children, him and I were best friends, but life happened, my family moved, his family relocated to the west coast, and my friend was lost. Until one day, using a search engine, we stumbled across each other's pages. Surprisingly, he lives only one hour away. If it wasn't for the site, I don't know if I would have found him again. (SocialNetworking2)

> There are times when a friend of yours or a family member that you hadn't spoken to in weeks, months or even years comes online and you can quickly give them a hello and a conversation of how each of you are can occur. Within this conversation you just brought back into your life a person you had "forgotten" about or "lost" just because you were both online at the same time. (InstantMessaging 4)

To peek into one's past and reconnect with an old friend can be a highly satisfying experience, the pleasure of which was mentioned by 44 percent of my interviewees.

The ability to better remain in contact with face-to-face friends and family was also highlighted, particularly among those who plan a geographic move or have recently moved:

It is really nice to keep in touch with those I have worked with especially those that are from places such as England, New Zealand, Australia and other places. . . . to know what they are doing in their life today. (Gamer2)

I recently moved 700 miles from home and IM has been a great way to keep in touch with friends and family. (Instant-Messaging2)

Having email has made me feel more adventurous in where I live—it's not so scary to move thousands of miles every few years (which I've done) because I know I can stay close to everyone I know on a regular basis. (WorkGroup1)

Some technologies provide creative or innovative ways of doing this:

my father is abroad for the year and we use VoIP to talk to him most of the time. my kids use the video function so they can see him and he can see them. it makes it easier. (ParentingForum1)

Online you can talk to your friends, and while you do that you can share photos with them or any articles you find on the internet . . . plus now you can put a camera, see them, talk to them with a microphone, you don't even need to type anymore . . . It does make life more interesting. (MusicLover2)

Utilizing technologies and social networks to interact and "meet up" with friends wherever and whenever we want has allowed the concept of friendship to be expanded and interpreted in new ways in portable communities (for more on this, see chapter 3).[3]

To take our communities with us wherever we go is an enormous benefit when we are literally on the move ourselves. Online social networks are very effective at helping us to bridge distances, because the pathways in these networks are always available, always "open":

When I am online, I am always checking up on friends, making plans, or keeping in touch with distant friends and relatives. More often than not, I am in interaction with at least one other person. . . . it is easy and convenient to keep in touch with people on a casual, day to day basis. (InstantMessaging1)

Portable technologies provide many ways for us to remain in this kind of continuous ambient contact: we can leave our computers on and remain logged in all the time, create text or IM away messages, or phone or text one another almost continuously simply to stay in touch or provide short updates. Our channels to one another can be "always on," always open, even as we move throughout physical space. This creates a kind of "nomadic intimacy," says sociologist Leopoldina Fortunati, in which "the public space is no longer a full itinerary, lived in all its aspects, stimuli and prospects, but is kept in the background of an itinerant 'cellular intimacy.' "[4]

In these frequent, often intimate, interactions, new people can be encountered and welcomed into our lives. Online acquaintances can become genuine friends:

> I've made a few new friends and although I haven't met some of them in person they're still people I talk to and enjoy having a friendship with. (MobileUser5)

> I blog to keep in touch with my friends and also make a few new friends here and there if they have similar interests to mine. (Blogger2)

And it seems to be a special pleasure to be able to contact people whom one might otherwise not have had the opportunity to meet:

> [My favorite site] is a great place to discover musical talent and I have got into bands that I would never have known otherwise. I like being able to actually be in contact with the members of the bands themselves, letting them know how much I like their music and it's exciting when they actually write back and take a real interest in what you have to say about them. (TVFan2)

> [Social networking sites are] fun but practical too because I have reminders of my friends' birthdays and a bulletin board of what's going on in the area each weekend . . . they're also the best way to hear about, keep up with, and communicate with the local bands in each area. I also visit in an attempt to keep those friendships. (MovieBuff)

Maintaining pages, profiles, and blogs on social networking sites is a popular way for organizations (schools, sports teams, businesses) to reach potential clients or followers, for celebrities (actors, musicians) to reach potential fans (and vice versa), and for any one of us to make new, un-expected friends.[5]

We also use online and mobile technologies to establish and expand local social networks in our face-to-face communities. Wellman and Hampton find that

> [e]mail allows neighbors to keep informed and in touch on their own time, without having to overcome the social and physical barriers necessary to knock on a door or do more than wave a hello from across the street. The result has been a local network of densely knit, specialized acquaintanceships with a relatively high frequency of interaction, both face-to-face and online.[6]

As SocialNetworking6 tells me:

> I can keep in touch with people that normally I would not have made the effort to keep in contact with.

People who use the mobile telephone, particularly for text messaging, report increased amounts and types of informal social contact as well. "With mobile telephony," sociologist and mobile communications specialist Rich Ling explains, "the key thing seems to be that it provides ready access to social networks for the individual and thus lowers the threshold for inclusion."[7]

It is common for people to use online and mobile social networks to strengthen or add to existing face-to-face social networks.[8] We tend to select most of our online and mobile "friends" from our own geographical areas or from among those whom we also know face-to-face. Portable technologies are often used to enhance communication among friends and family, make plans, and maintain social contact outside of day-to-day face-to-face conversations,[9] though there is some evidence to suggest that this is less likely among people who spend much of their time in chat rooms versus other forms of online media; chat users generally spend more time with others whom they will *only* meet online than, say, email users.[10]

With these networks pervasive, mobile, and always on, it is easy for individuals to "switch fluidly from network to network, using their communication media to contact the social network needed for each moment."[11] Barry Wellman and his research team call this *networked individualism*—the idea that each individual stands at the center of a web of connections and networks that he or she accesses and uses as needed. This type of social connectedness, Wellman argues, represents a transition from place-to-place to person-to-person connectivity. Moving around with a mobile phone, pager, or wireless Internet makes people less dependent on place. Because connections are to people and not to places, the technology affords shifting of work and community ties from linking people-in-places to linking people wherever they are. It is I-alone that is reachable wherever I am: at a house, hotel, office, freeway or mall. The person has become the portal.

For in online and mobile environments

[p]eople remain connected, but as individuals rather than being rooted in the home bases of work unit and household. Individuals switch rapidly between their social networks. Each person separately operates his networks to obtain information, collaboration, orders, support, sociability, and a sense of belonging.[12]

We can now obtain these kinds of things on an as-needed basis, accessing networks and communities easily and often via portable technology. This allows us to make and "juggle" many social connections, as numerous social ties and networks are established, offering almost limitless opportunities for social connectedness. But it places substantial responsibility on the shoulders of the individual—the "portal" for all this connectivity—to manage and coordinate these networks.

I favor another metaphor here for understanding the array of social connections each of us makes—what I call a *portfolio of social connections*. Modern individuals must manage a great number of social ties and communities, both face-to-face and sociomental. In my view, we each maintain a kind of mental portfolio that consists of all these varied connections. We use this as a kind of "planner" to organize and make decisions regarding all these opportunities and obligations. Research into the processes and conditions that establish sociability in portable communities and help us manage all our connections should continue so that modern individuals are well prepared to navigate these pathways successfully. For the traffic of sociation along online and mobile social networks will only increase.

CONVENIENCE

Much research demonstrates the sheer convenience of using portable technologies to develop social networks. We find wireless computers, mobile phones, and PDAs convenient and for the most part easy enough to use, and appreciate the flexibility they offer.[13] To be sure, many of us do not scratch the surface of the possible applications of all the technologies we use and could be using,[14] but we get by pretty well, and so turn to these technologies to accomplish a number of tasks that once took much more time and trouble. Like InfoGathering3, we appreciate the ability to "communicate with people much faster," more frequently, and more easily.[15]

Because these technologies can be accessed from anywhere and everywhere, at any time of the day or night, they provide special conveniences to the user:

> I IM with a friend who is up late like I am. I would feel bad to call her house at 11 or midnight, but if she is online, that is OK. (MobileUser3)

> Without email I would waste a lot of time playing phone tag with people when now they just send me an attachment with whatever they need. (Activist)

> At work it keeps me connected and gives me the short breaks I need to stay engaged in my work. (WorkGroup1)

> Text messages are sometimes clerical memos on when we'll be out of meetings or what to bring home. I love getting texts. They're short and sweet and someone took the time to key them in. (SocialNetworking1)

> I like the internet because people can respond when and how it fits into their schedule. (MessageBoards1)

Mobile phones and devices, in particular, have become indispensable in connecting with others frequently and conveniently because of their small size and portability.[16]

Portable communities generally have low "entry and exit costs"; compared to face-to-face communities, they are relatively easy to become part of and to leave.[17] This is a great boon to people looking to feel a part of a social network without the same responsibilities or obligations that face-to-face interaction carries:

> It's nice to have the option of writing at any time of day or night in an email, and knowing a friend can take the time to

respond whenever he has enough time and enough to say. (WorkGroup1)

I found chatting online an easy way to stay connected and I could easily get out of the conversation by saying I had homework to do or had to be somewhere and they would understand. (MobileUser1)

The use of a mobile device, Rich Ling points out, relaxes "the implicit contracts around time" and allows us to use time as we need to in order to accomplish what we wish.[18]

Portable technologies are rapidly becoming critical to the coordination of everyday activities, even changing the ways in which activities will take place:

I use email to coordinate family gatherings and dinners with friends and to "touch base" with family and friends. (OnlineStudent1)

My wife calls on the cell a lot to give me chores—to stop at CVS or Kings to get things on the way home. I also call my mother a lot on the road. (InfoGathering7)

It's much easier to make lunch plans or coordinate bridge games through email. I don't have to make lots of calls to arrange where we are going and when. (SupportGroup6)

This weekend I was at Final Four activities in downtown Indianapolis before a severe storm hit. Some friends and I got separated, and due to the large crowds and tornado sirens going off, we couldn't hear each other to find one another on the cell phone. Text messaging helped this situation and we were able to reunite. (Social Networking7)

Mobile phones now "invade routine behavior of all kinds," says sociologist of technology Hans Geser.[19] We have become accustomed to making or reconfiguring plans last minute or in transit via portable technologies. This inspires the near-continuous coordination of activities, or what Ling calls *microcoordination*.[20] Of the people whom I interviewed, 55 percent are *microcoordinators*—practiced in organizing social activities and events "on the fly."

It seems, as Ling observes, that portable technologies have resulted in time itself being "softened." We now tend to plan events more loosely, knowing that we can make changes at any time. Schedules are not always strictly adhered to; lateness is common. Portable technologies permit us to arrange and rearrange social events (and therefore time itself, in a way) on the go, even as events are happening.[21] This "softens the schedule," adding slack to the precise nature of time-based agreements. It is now almost unthinkable to return to a time when we did not microcoordinate activities in this way.[22]

With portable technologies, we can keep in touch with many people at once by sending single messages to groups of others simultaneously. We can use blogs for this purpose:

> I keep in touch with [friends and family] through online groups so that when something fun is coming up I can invite them. . . . I can mass-message people on [my social networking site] and posts in my blog. (Blogger2)

. . . mass emailing:

> i use emails for things that I want to send to a lot of people not just one. my emails usually consist of jokes, chain letters, cute pictures, etc. (Lurker1)

. . . and email, IM, and text messaging:

> When folks are too busy to talk directly to you they do manage to respond to emails and IMs. (OnlineStudent5)

> I use the email and my cell phone to help me stay in contact with my groups. The email is fast and convenient, and my groups usually reply back very quickly. . . . Doing this has made my life as far as staying in contact with my groups easier and less stressful. (InfoGathering5)

> IMing is something that allows me to stay in contact with a lot more of my friends and lazy as it may sound, it is a lot quicker than using the telephone (even a cell phone). IMing provides us (the users) to leave a message just through typing a message, tell fun stories at the time it happens, or just lets you know where everyone is, without possibly interrupting them with a phone call or even a text message. (InstantMessaging3)

Email and text "copying" allows us to send information and communicate easily with more than one person at once. Doing this facilitates interactions, solidifies group identities, and helps people develop alliances. It also allows us to perform multiple tasks in short periods of time.[23]

The convenience and efficiency of using portable technologies to interact are noted by adults and children alike as prime reasons for their popularity. Even the simple act of forwarding a joke or a picture to many people at once can be seen as a "bonding gesture" made possible by portable technology.[24] Permitting us to reach out to others reliably and efficiently, "the technology becomes a medium through which the social order is maintained," Ling notes. The social networks we form strengthen and firm up our societal infrastructure. And they become the means through which very personal kinds of connections are established, including those involving dating, romance, and sex.[25]

DATING, ROMANCE, AND SEX

As we have seen, it is easy for online and mobile relationships to take on a flirtatious, even intimate quality. Often, time spent in portable communities turns romantic or sexual in nature, whether by design (participation in a "matchmaking" website) or by chance. At least three of four internet users who are single and actively looking for a romantic partner have, at this writing, engaged in at least one romantically flavored activity online, from searching for information about prospective dates and local singles scenes, to flirting via email or text or instant messaging, to visiting dating sites.[26] Engaging in romantic and sexual activities are popular uses of online and mobile technologies.[27] Whether we find relationships that are continued offline or are kept in the online realm, many of us look to meet romantic and sexual needs as we travel the pathways of our portable social networks.

As discussed in chapter 4, time spent online or with mobile devices often has a playful, flirtatious dynamic. But deeper romantic and more overtly sexual activities are available in portable communities as well. These include romantic or sexy conversation, masturbation, and participation in phone sex or *cybersex*, which is the exchange of suggestive or explicit erotic messages or sexual fantasies with others who are usually online or using a mobile device at the same time. Some specifically seek sex partners; some look for dates. Those who seek partners for casual dating online do not seem to be particularly anxious about physical romance and do not seem to suffer excessively from low self-esteem, though they do tend to be somewhat more sexually permissive than the general population.[28]

Some romantic and sexual needs can be met with the help of portable technologies. It may be preferable or simply appealing to be able to size up potential dates or romantic partners from the comfort and protection of one's own home. Dating sites may be general or more specialized, catering to people looking for serious relationships or directed toward those seeking primarily sexual encounters.[29] Currently, some 11 percent of all internet users and 37 percent of those who are "single and looking" say they have visited dating websites. A majority of them say they have had positive experiences, and 17 percent have entered long-term relationships or married someone they met online.[30] Online social networks and social networking sites that assist people in finding romantic partners online have a success rate equal to, if not exceeding, that which would be found using face-to-face means. "Online dating can work for many people, leading to a successful meeting for almost everyone we surveyed," says Jeff Gavin.[31] In fact, traditional "matchmaking" services are increasingly being supplanted by the online variety.[32]

Dating in this way has no real-time demands, conversationally or physically, though people may choose to IM or meet synchronously in a chat room or at a live online event. It thus serves those with a hectic or busy lifestyle, as it provides an outlet for short-term comfort, excitement, and distraction.[33] Only four of my interview subjects disclosed having dated over the internet, possibly because this was not a major focus of my research and not something people generally volunteer unless, perhaps, when specifically asked (my questions were more general and thus not specifically about dating). At any rate, none of my interviewees reported finding major romantic success, although MobileUser1 said that several good friendships have resulted:

> When a single man would post that he liked my photo and wanted to chat online, it boosted my ego. Then when we would call one another on the phone it boosted the excitement. . . . I actually dated three or four of these men, they were all very nice . . . though none of these relationships went further than one or two dates. There is one exception, and he and I are the best of friends today, although have no love interest.

MobileUser1 goes on to say that

> Online groups have impacted my life in that I have met some very nice men to date as well as some not so nice. Without online dating services it would be much more difficult to meet

potential partners. . . . I also like to see their photo online so I know who I am meeting.

Online dating tends to works well because we reveal much about ourselves in online environments. We may even be able to sidestep confining, stereotypical roles in the process.[34]

After corresponding, exchanging photos, checking one another out, and possibly "meeting" via webcam, phone, or other internet or mobile device, those who feel they have made a palpable connection often move their romance offline.[35] It turns out that these relationships may have a *better* chance of working out than when the couple has first met by more conventional means. This is because the often quite personal revelations that have been shared and understandings that have been reached online give the relationship a kind of "head start." Without external distractions, couples can share information and evaluate their compatibility in a more leisurely and perhaps thoughtful manner. In addition, those who take time getting to know one another online before they meet face-to-face tend to have more successful relationships than those who do not. Internet-initiated relationships, once taken offline, are more likely to survive than those initiated face-to-face.[36]

The 89 online couples sociologist Andrea Baker studied all decided to meet offline after beginning their romances online. She found that, on the whole, couples that communicated longer before they physically met stayed together longer and formed permanent bonds in greater numbers than those whose communication was more short-lived. Additionally, couples who became involved in cybersex before they had ever met face to face tended to end their relationships somewhat more quickly than those who put off explicit sexual activity until they met face-to-face. Just as in physical space, moving slowly in an online relationship and prolonging its initial stages can be a good idea.[37]

When flirting, dating, and even sex take place in online and mobile spaces, we can find ourselves highly engrossed in the experience. "Cybercheating" on partners and spouses is on the rise. The anonymous, convenient, and occasionally escapist nature of the online experience combine to make infidelity of an emotional rather than strictly physical nature increasingly common in portable communities.[38] Individuals who are less satisfied with their face-to-face relationships seem most likely to seek additional support and intimacy online. In addition, those who are less satisfied with their offline sexual relationships may engage in more explicit cybersexual behaviors to "compensate themselves sexually."[39] Emotional or physical infidelity and other flirtatious, romantic, and sexually oriented behaviors can feel just as real and be just as meaningful as those that occur in face-to-face settings, with consequences that can be just as serious.[40]

Friendly or romantic behavior can be more ambiguous in nature when it takes place online than offline—although flirtatious conversations, early romantic experiences, and friendships with sexual overtones can be highly ambiguous in any setting. Considering the many ways in which friendships and relationships are now enacted and interpreted, emotions expert and professor of philosophy Aaron Ben-Ze'ev says that we are in a time of rapidly changing sexual norms. Eventually, he predicts, we will likely develop more flexible societal views toward sexuality, monogamy, and marriage. As I have written elsewhere in a critique of Ben-Ze'ev's work, it seems that sexual exclusivity itself

> may well be redefined in time, as people increasingly enter into these new kinds of ambiguous, are-they-or-aren't-they-sexual (or infidelitous, or adulterous) relationships, both online and offline. Some couples negotiate new "rules" for cyber-relationships—new meanings for "infidelity" and "adultery"—that may result in a new "romantic flexibility." . . . Depending on how these relationships are approached, defined, and play out in people's lives, they can have a range of effects—they can fill us with happiness and positive energy that might result in closer bonds with all of our loved ones, or anxiety and misery if they are uncontained, insufficiently understood, or indulged in compulsively.[41]

Portable technologies permit us to build and pursue numerous relationships of all kinds, almost simultaneously, and almost constantly. Some of these are likely to be difficult to define, especially as norms are now shifting to allow for greater flexibility in the definition of social relationships (and, for that matter, of communities themselves). Modern people are ever more frequently placed in uncertain situations with uncertain norms to guide us. We would benefit greatly, as individuals and as a society, from continued research into these new social arrangements and the role of technology in helping to shape them.

Disappointment and uncertainty can occur, of course, in all kinds of social interactions. But many positive romantic experiences are found and long-term commitments made in relationships initiated in portable communities. Though most of my interviewees did not report online dating experience, it is clearly becoming a common and convenient way for people to find one another for all kinds of friendly, romantic, and sexual relationships. Research into the scope of these activities and the new norms that will inevitably result will be much needed as well. We must also consider how portable technologies permit us to utilize our social networks to accomplish any number of instrumental tasks, including the gathering of information, learning, and working.

LEARNING, WORKING, AND GETTING THINGS DONE

Easily accessed portable social networks provide an excellent means to get all sorts of things done. From gathering information to going to school, from working to shopping to learning about absolutely anything, it seems that no matter what we do, we turn to portable technologies to help us do it. Fully 97 percent of my interview subjects mentioned using online or mobile technology for some practical, useful purpose. According to InfoGathering5:

> Participating in online groups . . . has allowed me to commu-
> nicate well with others, listening to their beliefs and views on
> things . . . to not only be more open to others and what they
> think but also to be more informed about worldly matters.

And InfoGathering2 says that in her portable communities, she

> learned how to communicate, be patient, how to accept and
> deal with challenges, and how to say "no."

We use these communities for an almost unlimited number of instrumental purposes, including, but not limited to, those mentioned most often by my interview subjects: learning, information gathering, commerce, scientific and academic inquiry, and work.[42]

We can now use online and mobile technologies to locate a piece of information at the exact moment when we feel that we most need it. Information has become a "primary good" in society, observes communication and media scholar Jan van Dijk. This means that "a particular minimum of it is necessary to participate" in society.[43] Skill in accessing information is also critical. Upwards of 40 percent of American internet users, as I write this, use search engines every day to access information of all kinds.[44] My interviewees report using their online and mobile social networks for almost every purpose imaginable:

> Participation in online communities has become unavoidable
> when I do research and search for things on the web (job-related,
> shopping tips, apartment hunting, etc.) (InfoGathering4)

I belong to [a dieting website] and visit it every day. It is helping me lose weight in a healthy fashion. I joined the website
because it is much cheaper than going to meetings . . . and if
we fail, nobody knows. (E-dieter)

[On the message board] members post questions regarding
buying tickets, getting directions to a gig, where to find the
best price for merchandise, and other members gladly offer their
assistance. (MusicLover1)

In portable communities, people both seek and provide information to one
another in massive amounts:[45]

There are so many different boards out there and so many
people asking so many questions. It is a great tool for people
to be connected like this and be able to get answers to questions. It's nice to be able to go to my main site and ask for
example, "I am having trouble narrowing down a squeak in my
front driver side axel, has anyone had this happen or know of
a common problem"—then within minutes I will have some
guy's suggestions or sometimes you get someone that has had
a similar situation, that they had fixed and they can enlighten
me to help me not go through tedious tasks, and get right to
the problem. There are so many resources and I hope people
take advantage of them. (SocialNetworking12)

This listserv has been a true life saver. It gave me an invaluable link to people who were coping with the same cancer as
I. It is a valuable source of knowledge and personal perspectives that I would not have any other access to. . . . Even my
physicians are impressed with the timely information I receive.
(SupportGroup2)

I get a lot of information about and hear rumors about when
my favorite band will be touring and get info on the best way
to get good seats for concerts. I hear about when new albums
will be coming out. (MusicLover4)

It is helpful for people who are new to a rare hobby to get
information and connect with people with more experience and
can tell them where to get books, supplies, etc. Even experienced
people can learn from one another. (Chatrooms2)

In collaborating this way, we simultaneously inform one another and tighten our social networks. As WorkGroup7 puts it, "It is very satisfying to provide info to other individuals with common goals."

Many portable communities are explicitly learning oriented. Whether taking a class online or learning in less structured way, people are becoming accustomed to using portable technologies to pursue knowledge and information anywhere, anytime—even when "on the go" themselves:

> As I was learning, I would just ask questions about what I wanted to know, and within a few minutes people would respond with the answer. That was probably one of the best aspects of [belonging to the group], instant information from qualified people. (WorkGroup5)

> [My favorite online group] is a chance for me to learn something and to interact with others. It also sharpens my research skills as well as my arguing skills. (InfoGathering6)

From message boards to wikis to social networking sites to search engines, there is no shortage of places to go to get answers to one's questions, no matter how quickly they are needed (or how impatient we are in desiring them):

> I feel that I am better informed from the information I get through websites. They also make random discussions more interesting if you know or have heard about something that someone else may have an interest in. (MobileUser7)

> I enjoy visiting news and information sites. They allow me to exercise my mind, talk to people who share my views or oppose them, and learn from all kinds of people from all walks of life. (InfoGathering5)

My interview subjects told me that the exchange of information in portable communities when and as they need it most has become a primary use of these networks.

Portable technologies are also in many ways revolutionizing formal education. Online education is becoming an increasingly popular way to deliver and access course content. Students can bring many educational offerings and information with them wherever they go. Educator Bryan Alexander says that college students can now "turn 'nomad,' carrying conversations and thinking across campus spaces, as always, but now with the ability to google a professor's term, upload a comment to a class board, and check for updates to today's third assignment—all while striding across the quad."[46] When and where such technologies are available, students can become "portable learners," producing and consuming knowledge no matter where they are.

Students can use portable technologies to become more critical learners as well. With numerous resources at their fingertips they can develop a deeper, more focused, approach to learning. The use of participatory technological tools such as blogs has been found to enrich student learning and knowledge and to correlate positively with learning outcomes in a way that traditional coursework does not.[47] Those who teach online or use technology in teaching report benefits as well, especially when becoming actively involved in the experience and establishing trusting relationships with students in structured yet flexible environments.[48] Educational activities carried out with the help of portable technologies have proven highly adaptable to instructors' and students' needs and have been linked to higher grades and greater student satisfaction and motivation among both children[49] and adults.[50]

Educational and learning groups function as full-fledged communities when members gain a feeling of belonging and purpose, share understandings, and develop an image of themselves, an identity, as a group.[51] Properly designed and moderated, they can also function, very deliberately, as social networks: members work collaboratively, exchange information, advice, and social support. And these effects can persist over time:

> Participating in online groups such as having classes online has enriched my life. I have met lifelong friends that I still connect with to this day, and it has changed the way I view certain issues. (Chatrooms1)

> Participation in an online group has allowed me to obtain my undergraduate degree and is now allowing me to pursue my master's degree . . . and helps me to feel part of the college community. (OnlineStudent1)

As we have seen, the lack of visual cues can encourage disinhibition, which can help us become even more deeply involved in portable learning communities than in the traditional kind:

> I join discussion groups that my classes have on the board. It is a cool thing and a cool feeling. Through this board, I have no fear expressing myself, and I am not afraid to write back to

someone saying, "I disagree with you." Whereas in class, I would feel uncomfortable to say it that way. (InfoGathering2)

Many of us feel more comfortable or less fearful opening up to others from a distance. Challenging questions can more easily be asked and considered; challenging goals may more easily be set and met. Portable communities can be "safe interaction spaces" for students.[52]

We also use these technologies to create work spaces. The internet and, increasingly, mobile technologies, have become integral if not indispensable tools for transacting business and commerce. At least one-third of internet users use the internet in the course of their work.[53] "Distributed work groups," "online workspaces" and "knowledge networks"—online communities of practice and their mobile equivalents—are now routinely created and generally quite successful in helping spatially separated people work, learn, and accomplish tasks together.[54] It is interesting to note that when younger people enter modern workplaces they are often much more technologically skilled than veterans. They have also been prepared for the work world differently than prior generations: they tend to excel at gathering information quickly (though not necessarily at assessing the credibility of this information), at multitasking and communicating with more than one person simultaneously, and at completing discrete tasks.[55]

In addition, blogging has become an excellent way for people to professionally interact in a productive, up-to-the-minute, yet informal way. As media and cyberculture scholar David Silver explains on his own academic blog:

My blog affords me academic connections to colleagues, students, and scholarly, artistic, and activist communities. Sometimes, the connections produce professional opportunities such as presentations, publications, and collaborations. Sometimes, the connections produce intellectual opportunities like multi-authored comment threads that further nuance an issue, idea, or interpretation. I enjoy and benefit academically from such connections, and . . . use my blog to share my research with my students and to share my students' work with my colleagues.[56]

Silver illustrates just some of the many ways we can use these platforms to learn about one another's work and interests and perhaps learn from one another and pursue interests *collaboratively*.

Many use their online and mobile networks to assist in job searching . . .

[My online community] has helped me find meaningful work opportunities. I can't imagine my life without this sort of resource. (WorkGroup1)

I visit job boards both in job hunts for myself and also in my line of business. I work as a recruiter so the bulk of my time is spent searching resumes. It really isn't appealing, but it is what I need to do. (WorkGroup4)

. . . to do their jobs more efficiently or effectively . . .

I am an active member of our listserv at work. We communicate by using it quite often and I find it a quick and efficient means of communication to a larger group of people as long as they read the emails that they receive. (WorkGroup3)

I am part of a few online groups, primarily business-related. In my current professional role, I work for a global organization and I participate in many email discussions/threads that go back and forth within my client groups. Although my interaction is constant with them, I have never met a large number of them face to face. For me, logging on to my email account makes me feel "connected" to my client groups, with the ability to quickly interact and "chat" with the click of a mouse. Therefore, my online group is more than just practical, but necessary. Today, with more and more businesses in global environments, the need to create such online communities is incredible. I have observed that those that are not technology "savvy" have difficulty working in such a modern environment. (WorkGroup2)

. . . and to escape from work (or *at* work) as well:

When I'm bored at work, I certainly feel like checking my email constantly. . . . In fact, knowing I can email with friends makes me feel less "trapped" at work—I have access to everyone I know, so I don't need to feel trapped or isolated and therefore can better focus on my work. (WorkGroup1)

Portable technologies are often used to bring us to a sociomental space that differs from our physical place, a benefit often enjoyed by those in work spaces they would prefer to (at least temporarily) escape!

There is a downside to consider as well. It is unclear exactly how helpful the internet is in helping people find jobs; research seems to indicate that it may afford a small advantage in that some other job searchers are not using it, but it is by no means an optimal way to obtain many types of jobs.[57] Some students may not do well with online education. And the credibility of online information must always be carefully assessed, for misinformation abounds on online and mobile networks:

> On my professional practice, I have had a painful experience to have reached a wrong conclusion based on misleading online information sources (or sometimes a hoax). It gave me and my company a financial damage (and my boss was not very happy, as you can guess). (InfoGathering4)

Unintended, sometimes unwelcome, consequences must always be considered when assessing the impact of technology.

Portable communities are often social and economic at the same time. At least one-third of American internet users shop on the internet, a percentage sure to rise appreciably. Though some of us worry about the riskiness of providing personal and financial information online, we tend to move forward with such transactions anyway when we trust the reputation of the person or institution with which we are doing business or there is a reward for doing so (like getting a good price!). And many of us simply appreciate the convenience, speed, and value of online shopping:[58]

> There are great ways to buy merchandise or products through a message board classified system, I have bought numerous used or new parts off of people that have been selling stuff on the board, usually at a great deal. (SocialNetworking12)

> When I get an email on a sale or a sale circular, I can be involved clicking on links and before I know it 30 precious minutes have gone by in the blink of an eye. It eats time, more than television, because at least TV has commercials. Internet is interactive and you are totally involved. (MobileUser3)

> Anytime I need to buy a product that costs more than $200, [my community] is one of the best ways to gain the true opinion toward that product. . . . On the other hand, the chances to receive distorted or misleading information has also been increased. (InfoGathering4)

Money need not exchange hands:

> We swap books through the regular mail, and sometimes gifts
> and postcards. It is interesting to learn about another culture.
> (ParentingForum1)

And MobileUser7 reports this success story:

> The other day I saw a chair on preview at a local auction house.
> I knew it was something good but I didn't know how good. I
> sent a photo with my cell phone to an expert in arts and crafts
> style furniture and he was able to tell me exactly what the chair
> was and a rough estimate of what it was worth. I bought the
> chair for $25 and sold it the next weekend for $700. I would
> never have purchased the chair in the first place without being
> able to use the technology I had to do on the spot research.

The popularity of eBay and other online auction and "swap" sites is further evidence of the willingness of many to make commercial transactions
online.

Online commerce definitely has a communal dimension. When we
rate products, places, services, and sellers online, provide opinions or commentary with regard to them, or compile lists of similar or favorite items,
we go beyond the sales transaction to form a kind of social connection.
There is much opinion and information sharing online; some sites specialize
in facilitating it (at this writing, these include Epinions or TripAdvisor). As
we so often do online, we want to share our thoughts and be understood,
to speak and be heard, to support one another, and, perhaps, to gain a
kind of status and identity by successfully doing so. Portable technologies
provide nearly unlimited opportunities to do so.

We even see the creation of a "herd effect" online. On online auction
sites, we tend to be willing to pay more for an item if others express an
interest in it.[59] This demonstrates, again, the sociability of the community.
Clearly, in ways both subtle and overt, we are aware of and responsive
to one another's behavior, both online and offline. This awareness, this
willingness to be of help to one another, be in contact, and follow one
another's recommendations, is clearly paramount in portable communities,
and is often really the point of it all—the "latent function" of the group,
as sociologist Robert Merton would put it.[60]

These networks that circulate throughout portable communities are
mental networks with mental pathways, and anyone with technological access
can use them to look for others with whom to form social ties and establish

community. The potential strength of these ties is sometimes debated. In fact, both strong and weak ties—and everything in between—are found in online networks.[61] The closest of relationships are built and sustained via portable technologies, while more fleeting, ephemeral ties are much in evidence as well. Most communities have a healthy supply of both. And most individuals manage portfolios of social connections that contain hundreds of social ties—weak and strong, face-to-face and sociomental.[62]

Even when weak, seemingly insignificant social ties are created, they can still have great utility. Sociologist Mark Granovetter has famously taught that weak ties create strong societies. They bring into contact people who might otherwise have no way to know of one another at all, thereby opening up pathways which eventually provide *all* members of one social network with access to *all* the members of a second network.[63] More recently, Caroline Haythornthwaite has examined how the use of applications such as email, chat, and telephoning can increase the strength of ties in networks and embed people in webs of connectivity. We obtain social capital—valuable resources that can pay off in critical information, such as job contacts—when weaker social ties "bridge" networks and communities together. In contrast, we gain intimacy and closeness when stronger social ties "bond" us to one another and keep our communities internally strong. And like our social ties, social networks can increase or decrease in strength over time. When their strength increases, the number of relationships they can support increases, as does the potential strength and emotionality of those relationships.[64]

Online and offline, this is how social networks work: by creating pathways and opportunities by which people in one group are exposed to people in another group and go on to create communities together. These networks are used for business or personal purposes, for sociability, work, learning, commerce—for the full range of motivations for social association. Modern societies are dense with these criss-crossing pathways and networks; knitted together of strong, weak, intermediate strength, and even latent ties (those which have the potential to be but have not yet been actualized).[65] And as these ties and networks connect us, the infrastructure of a society is built.

Social ties, networks, and communities exist in both face-to-face and sociomental forms. Though not always visible or in our conscious awareness, those that are sociomental and created in technological use reliably, and almost constantly, bring us together. So far in this book we have looked at many of the social dynamics inherent in portable communities: the cognitive infrastructure that makes these social ties and networks possible; the emotionality, intimacy, fun, and playfulness that flow along their pathways; the gathering of information; the learning that takes place;

the work that is done. In the next section of the book, we widen our perspective on online and mobile connectedness to consider the external dynamics and social implications of portable communities. We will focus on the resonance between portable communities and the world in which they are situated, and consider their impact on our societies, our relationships, and our selves.

III

EXTERNAL DYNAMICS
The Portable Community in the Society

6

BEING THERE

Constant Availability

From those on our cell phone contact lists to our IM buddies, from those who visit our websites and social networking spaces to those who respond to our blogs or posts, people now effectively surround us at all times. It is as though they are always "out there" (or "in here"—in our pockets, like our phones) and that we are in perpetual contact with them.[1] Chatrooms1 observes that "It doesn't matter what time, someone is always online"—or on the other end of a mobile phone or device. Though we might not always be able to make immediate contact with the *exact* person we're trying to reach at the *exact* time we want, the likelihood that we can rather expeditiously contact *someone* in one of our communities—someone with whom we have *something* in common—is unprecedentedly high.

There are undeniable satisfactions associated with having easy and near constant contact with a large number of somewhat like-minded others. Being able to interact with those with whom we have something in common, at our convenience, can be extremely rewarding.[2] We may be able to get valuable information, social support, or just a kind word, at the moment we need it most. We may be less fearful, less anxious, believing that we are safely connected in the event of an emergency. In a larger, more diffuse sense, we can obtain the sense that others who are a lot like us are "out there" in the world—people whom we can contact at any time of the day or night if we so desire. This can help us feel more emotionally connected to others and the world around us—less alone, more "plugged in" to society in general. Blogger1 reinforces this:

> I think that many people feel it's important to have a connection
> with others all the time.

To have a community of like-minded others available at your disposal, wherever you go, whenever you need it, can be useful and comforting.

But constant availability to others can also be a source of anxiety. We may feel that others' eyes are always on us, checking up on us. We can feel a claustrophobic sense of having no space, no time to ourselves.[3] There also are definite threats to our privacy when so much of what we do is publicly accessible. As SupportGroup5 says:

The pro side is I'm available, and that is the down side, also.

This chapter looks at both the upside and the downside of always "being there"—at both the beneficial and harmful aspects of our being so very available to one another in a technological society. It examines the comfort and companionship we derive in portable communities, the way that we now envision and respond to emergencies, anxieties, apprehensions, and technological overload, and the considerable impact of all this technological connecting on privacy.

COMFORT AND COMPANIONSHIP

One of the most common comments I received from my interview subjects was that they felt that their portable communities, and a good many of the people in them, were always available:

I tune into the online world via my Blackberry when I am at work or home all the time. We have online groups that I communicate with my co-workers through and it keeps us connected and all lines of communication open. (MobileUser8)

When I am at school, I am instant messaging all the time. . . . I can contact pretty much anyone I know within minutes. We are able to keep in constant contact if we choose. . . . Several days, I may find myself on the computer all day. (InstantMessaging1)

I sit in chat rooms all the time, and communicate with people I do and do not know on a consistent basis . . . (Chatrooms1)

I can get in contact with almost any of my friends at any time of the day wherever I am or wherever they are . . . if someone needs or wants to get in contact with me they can do so at any point in the day. (SocialNetworking10)

. . . or, if not constantly, at least "frequently":

I look forward to checking my email throughout the day, and often exchange a series of short emails with friends throughout the day. (WorkGroup1)

I like to receive little updates or five-minute chats from old friends or even current friends who I have lost touch with because of hectic schedules. I also enjoy receiving short romantic messages from my girlfriend when we are apart. (InstantMessaging1)

We tend to check email, instant messages, and text and phone messages so often because it is so easy to do and so instrumental that it becomes a habit. But in a more latent sense, doing so helps us feel the presence of others; it helps us internally "measure," in a sense, the strength and continuity of our social connectedness.

Human beings derive great comfort from knowing that others are around, are "there" for them. Even when we must be physically separated from certain other people, it is comforting to have the feeling, the "proof," that they are still there, still in our lives. As discussed earlier, the ability to provide people with near-constant background awareness of one another—what is called ambient copresence—is a feature of many portable communities. People who text or instant message can leave away messages for one another so that even when they are not reachable they are still communicating, providing details about where they may be and when they may again be available. As InstantMessaging1 put it,

As soon as I get home from being away or out, I tend to check my computer to see what other people are doing by reading their away messages or see if anyone IMed me. Text messaging is great too for little messages like "Hey, I was thinking about you" or "I'll be home at such and such time."

To maintain this constant, ambient sense of one another not only increases the contact and communication among us but sets up an *always-on mode* in which communication channels are always open and others always seem to be around. This makes us more continuously aware of one another and provides "a sense of connectedness, bringing people together."[4]

As portable technologies are so easily brought with us wherever we go, people now "pop in" on one another frequently. It has become normative, especially for the younger among us, to use mobile phones or handheld devices to check in with one another very frequently, simply to announce one's presence, provide a brief update as to casual goings-on, gossip, send a quick photo, "kill" a few minutes, or "just say hi." Mobile phone services (such as Twitter) now allow users to share with great frequency the smallest and most mundane details of their days with an audience (as in "I just had breakfast," "I'm sitting on the porch now," or "I'm listening to Springsteen"). As Daniel Graf, the founder of Kyte, another mobile social networking service, says, "now you can share your life over a mobile phone and someone is always connected, watching."[5]

It might seem excessive, even unpleasant, to receive constant updates of your friends' goings-on. But social technology writer David Weinberger provides another view on his blog, as he shares that

> [a] lot of it of course I don't care about. But it turns out that I do like hearing that . . . an Italian friend I see every couple of years is sitting on his porch, drinking wine and watching the sunset. I do like hearing that [another friend], who I unfortunately run into very rarely, is working on a presentation to libraries, which she then shares with her Twitter pals. I do care that . . . a Canadian musician I've only met once, is rehearsing for a live show. This is, to mangle Linda Stone's phrase, continuous partial friendship, and it's a welcome addition to the infrequent, intermittent friendships we're able to manage in the real world.[6]

Technology consultant Linda Stone calls the practice of constantly giving little bits of attention to everything around us *continuous partial attention*; it can be thought of as an effort to "not to miss anything . . . an always-on, anywhere, anytime, anyplace behavior that involves an artificial sense of constant crisis" that derives from our desire "to be connected, to be alive, to be recognized, and to matter."[7] Paying attention to the world in this way can be simultaneously functional and stressful, as we shall see.

Even in less extreme cases, though, it seems clear that we are now making "additional communication we might not have made before," communication researcher Leslie Haddon notes.[8] It is not necessarily the messages that matter most, but the gesture of staying in touch, of remaining connected, nearly all the time. For young people, journalist Betsy Israel finds, "to be 'on' with your friends is a birthright."[9] Groups of people can stay in contact with one another in this way all day long, all the time, if they wish. InstantMessaging4 tells me that

> Everyday if I am not already signed in online I make sure that I do, solely for the purpose of seeing what everyone is doing or just to say a quick hello. . . . Sometimes there are moments when I am online and no one is around (they all have away

messages up), but I'll usually still leave them a message and just say hello, or my main purpose will just be to find out what they are doing . . .

Communicating so regularly strengthens in-group relationships and allows members to subtly (or not-so-subtly) monitor one another's presence.[10]

In this fairly constant (if intermittent) stream of text messages, IMs, photos, and other short communications, often made without having anything special or particular to say, we are given to sense others' presence and we can assume that our own presence is being sensed in return. Communication and media researchers Mizuko Ito and Daisuke Okabe state that

> [u]nlike voice calls, which are generally point-to-point and engrossing, messaging can be a way of maintaining ongoing background awareness of others, and of keeping multiple channels of communication open. This is like keeping instant message channels open in the background while going about one's work . . . (the rhythms of which) fluctuate between what we have characterized as chat and a more light-hearted awareness of the connection with others through the online space.[11]

This "light-hearted awareness" actually serves a serious purpose: it helps us confirm, as frequently as we wish—even perpetually—that certain relationships persist. It can provide us with a feeling of constant companionship that can be welcome in a stressful environment, or in a strange or lonely one.[12]

The idea that our friends, family, and acquaintances are always "out there" and can be reached fairly easily can comfort us on a very deep, tacit level.[13] "Just as the individual's deprivation of relationship with his significant others will plunge him into anomie," sociologists Peter Berger and Hansfried Kellner explain, "so their *continued presence* will sustain for him that *nomos* by which he can feel at home in the world."[14] We need to feel at home in the world, to feel that we are not alone. Even just having our portable technologies with us or *on* us can give us these comforting feelings:

> When I don't have my cell phone with me, I rush back to my room to get it. I don't really always use it, I just feel happy to have it with me. (AnimeLover)

> The internet for me is always on but I am not always around it. . . . I am not glued to my computer or IM, I just enjoy

knowing that I could be online and have the ability to talk to my friends whenever I want. (InstantMessaging4)

As InstantMessaging4 puts it so well, simply knowing that we *could* be in touch with others at nearly any time provides a sense of satisfaction. SocialNetworking1 feels that

> [t]here's something very comforting about knowing that there are little pockets of "experts" or "regular folks" who are always available to dispense information.

And it can provide a needed sense of connectedness when one is away from home, as Chatrooms2 and WorkGroup6 describe:

> I take my laptop whenever I travel . . . so I can feel more connected to familiar things when I travel . . . it makes me feel still connected with my job or friends or family back home.

> When I was away I kept in touch with my son via text messaging. Knowing I was in touch with family was comforting while I was out of the country.

We feel less alone this way, for we have companions, family, or friends available any time we want or need them. In fact, when people feel their strongest need to stay in touch with others, they often use technological rather than face-to-face forms of communication to do so, because technology connects people so reliably. Members of a portable community are, quite simply, reached more easily, more often, and more reliably, than are many members of traditional communities.[15]

This greatly enhances our sense of connectedness with others and our sense of being in a genuine community with them:

> I feel more connected with my friends and family knowing that they can reach me at any time. (MostlyEmail2)

> My kids keep in touch with me during the day (via text message) which does make my day pleasant. (WorkGroup6)

> I carry my phone with me at all times. I feel empowered to be able to connect with whoever I need instantly. . . . I also have a much greater sense of comfort knowing that I can drop a

line to the listserv and within minutes begin to receive replies. (SupportGroup1)

Many told me that being part of a portable community simply helped them feel that they are more generally "in touch":

It really just gives me a feeling of being more in touch with people, like they can reach me whenever. (SocialNetworking5)

I like the "in touch" feeling . . . (MobileUser5)

It is both comforting and empowering to feel that you can access others at any time, a prime reason why mobile phones have become so ubiquitous.[16]

Constant access to others can also serve more instrumental functions. Parents can now more easily supervise children from a distance. The mobile phone serves as a kind of "umbilical cord" through which parents can ascertain their children's whereabouts (adults also use it to check up on one another, as I shall soon discuss).[17] Doctors can now be more continuously in touch with patients, managers and other professionals can be available around the clock, and people can take on all kinds of "pervasive roles" where they are expected to be always on duty:[18]

I have to be in touch all the time for my job. As a consultant I have my laptop with me at all times, as well as a cell phone, and will be getting a blackberry pretty soon. Most of the time it doesn't bother me, it's actually comforting to know that you can get in contact with someone. Other times it's hard to relax when you know you might be called upon. One good example was a few weeks ago. I was pretty much sleeping, and my fiancé heard my cell phone vibrating and answered it. It was the San Francisco office calling to see if I could fly out in the morning. Needless to say, I was on the 7 a.m. flight the next day. It's times like that that cell phones shouldn't exist . . . (WorkGroup5)

Certainly, constant availability can disrupt one's comfort or schedule.

But in a profound way, portable communities help their members feel "plugged in," not just to one another or to their social network but to something even larger—to *society*. A person who participates in them can minimize the isolating, perhaps even terrifying, feeling that one is alone, cut off from the rest of the world.[19] This is a phenomenon perhaps best demonstrated by its inverse: when we are temporarily *out* of touch with others, as during a power blackout or a technology-free vacation—or even when the phone isn't charged or the wireless isn't working—we can feel wildly off-kilter and out of place.[20] As Kate Fox notes:

> Much has been written about the loneliness, isolation and alien-ation of modern urban life, but few commentators have noted the important role of the mobile as an antidote to these evils. You may be surrounded by uncaring strangers in a busy city street, or working in a competitive, unfriendly office, but your mobile gives you a lifeline connection to your own social world, your village green, your garden fence. Carrying your social support network in your pocket, you'll never walk alone.[21]

Now, people can get their network, their community "to go"; they can for all practical purposes put it in their pocket and take it with them everywhere. This can provide a sense of companionship unparalleled in its constancy. And it can be especially beneficial—and perhaps just as significantly, *perceived* to be beneficial—in the case of the emergency.

EMERGENCIES

The ability to reach others more quickly in an emergency is often touted as one of the biggest benefits of portable technology and mobile phone use in particular. It is one of the main justifications given for obtaining a mobile phone in the first place, although people usually begin using it for many other things soon thereafter.[22] "The notion of a mobile telephone as a lifeline is one of its most central images," communication and mobile technology researcher Manuel Castells and his colleagues point out.[23]

Keeping a phone or other portable device nearby can help people feel safer during ordinary situations. It can give them peace of mind that if they were to be in acute need of assistance (such as having car trouble or being in an accident), or stuck in a remote location, they would remain safe and could get into immediate contact with others. Portable technologies are also often used when there is a chronic, ongoing need for contact, such as when one is ill or physically separated long-term from someone with whom they wish to remain in contact.[24] Underlying the casual use of mobile devices is the sense that we carry with us a lifeline: if we need others, they are reachable.

Mobile phones and portable devices are also expected to keep us safer during extraordinary situations. They are used to let family members and loved ones know that one is in peril and to help them check on the status of loved ones, as happened during the September 11th, 2001 terrorist attacks. Many rescue and information-dissemination efforts are now coordinated by portable device. Private citizens can mobilize quickly as "first responders" to emergency sites. Mobile phones are also indispensable in truly desperate situations as people use them to say their final goodbyes, to reach out one last time. Again, at such times, the image of such devices as a "lifeline" is well drawn. But mobile phones are "deathlines" as well. The September 11th airplane hijacks, and other crimes and attacks, were coordinated by portable devices. Mobile communication can not provide absolute security to anyone; it can be used to destroy as well as to rescue.[25]

Some of us—women in particular—are known to use the mobile phone as a kind of "shield" to fend off unwanted face-to-face interaction, potential harassment, or dire or simply unpleasant situations. We may go through the gestures of pretending to talk on the phone to signal to an outside agent that we are in contact with someone and therefore somewhat protected.[26] One woman interviewed by Rich Ling used this strategy in a situation that she perceived to be an emergency:

> I was alone in town once and there was a car that followed me and suddenly it screeched to a halt. And so I immediately took out my mobile and called my boyfriend, and they (in the car) left immediately when they saw the mobile.

Two of Ling's other interviewees describe how merely going through the gestures of calling provides a feeling of safety:

> When I am alone, often I just call someone just to talk, if you know what I mean. If I am going to go home and it is late at night, then I often call someone to talk to them while I walk.

> If I am alone in town, for example, I get out my phone and just start talking.[27]

Used this way, the phone can be used to call for help, or it can simply be a "prop or symbolic icon for connectedness," says Ling. It provides "vicarious protection . . . it is being used to communicate the idea to potential perpetrators that the woman is in touch with others and thus is not 'alone.' "[28] Not only do portable technologies keep us socially connected, they signal to the world that we are connected—a perception of connectedness that may at times be just as important to us as the connections themselves.

We now tend to use portable devices not only to contact someone in the case of emergencies but to express immediate thoughts and feelings as though they fit into the category of the emergency. We want to share what something is like for us at the moment it occurs. We can share current feelings and states of mind in all of their immediacy (calling to say "I got the job!" or "I love you!" or just "I'm bored—what's up?"). The ability to share so many thoughts and emotions with such immediacy illustrates well the idea that we are "always together"—even as some worry that in an avalanche of quick, breathless communiqués, depth and complexity of communication may suffer along with our attention spans.[29]

We have now come to expect to reach each other at a moment's notice:

> I need to know that I am available to people when they need me. (WrestlingFan)

> I like knowing that I can contact anyone I want whenever I want, and if someone really needs to get ahold of me they easily can. (WorkGroup1)

> It is nice to be available at all times because if there is an emergency you know right away. (Gamer2)

As MobileUser6 puts it:

> My cell phone gives me a feeling of comfort to know that if I am needed, whether for an emergency or someone just needs someone to talk to, *I am there*. (emphasis added)

Often, it is the *feeling* of safety that we seek:

> Most of the time, I carry my cell phone "just to be safe." It's like a security blanket if my car breaks down, or if I get lost, or if there is some kind of accident, or if I am running late, it offers comfort to know that whatever I'm facing, I don't have to be alone. (MovieBuff)

> My phone makes me feel safer because of the fact that I can call for help anytime. (MobileUser5)

> Cell phones provide a feeling of safety, knowing that if I need to reach someone, for example in an emergency, that I will be able to reach them. (WorkGroup2)

The desire for safety, then, intersects with the desire for social connectedness. Although we know, intellectually, that a serious emergency is unlikely to materialize at any given moment, we still look to others to help us feel safe and comforted, believing we can reach them efficiently and expeditiously, or make people think that we can, whether or not a legitimate emergency ever rears its head.

Portable technologies and the communities that coalesce around them can indeed help us feel safe. Having people always at hand, always reachable, can be comforting and satisfying. But in exchange for this sense of safety and comfort we may be trading our tolerance for disconnection. Gamer2 admits that

> [i]t affects me because if I am always available, I expect other people to always be available. I have lost some patience with waiting online or calling someone, if they don't answer I get upset. When I'm "out of touch" sometimes it feels nice but I feel anxious because what if someone does call and I miss it. (Gamer2)

> I know for me it is hard sometimes not being around a computer because you are wondering what your friends are doing or wondering if someone you haven't spoken to in awhile has IMed you . . . (InstantMessaging4)

When technology and communication are pervasive, it becomes easy to recast practically every request or desire as an emergency:

> I always have my cell phone on me. It's reassuring to know it's unlikely that I'll be stuck somewhere with no way to contact anyone but when I find myself without it it's like I panic. I think, "oh no I forgot my phone at home, what if something happens? What if someone needs to get ahold of me in the next few hours?" It takes a few minutes for me to realize how unlikely it is that someone will not be able to wait for me for a couple of hours. (MovieBuff)

There is, then, a discomforting flip side to the comfort of constant availability. Stress and anxiety can result when we happen to be disconnected or out of touch. Even just to worry that we *may* be out of touch someday, sometime, can be fearful. In a deep, almost existential way, we can begin to feel the effects of being *too* in touch with others.

ANXIETY, APPREHENSION, AND OVERLOAD

Technology anchors us in our spaces and relationships. We have become so used to constant connectedness and availability that it feels all the more uncomfortable when we are, temporarily, out of touch. We have come to depend on portable technologies to feel anchored, everywhere and all the time; without them, we feel "lost." The anxiety that this produces can be profound:[30]

> If I did not have my cell phone with me, I would feel lost. (Lurker4)

> If I could not use my cell phone and computer, I would be totally lost. (MobileUser3)

> I do feel lost without my cell phone because I'm so used to it being always on me. (Lurker1)

> When you are somewhere and there is no access to a phone or no way for someone to reach you . . . I think many people feel out of place. (Blogger1)

The metaphor of feeling somehow physically "lost" when out of touch technologically is not accidental. Portable technologies anchor us in space and give us such a definite sense of place. It is, in part, that sense of place that is missed in their absence. This is all the more interesting when one considers that we are always somewhere, in some physical place, when we use technology. Still, we feel lost when we are denied sociomental connectedness because it is so important to us to be cognitively connected; to feel firmly "in" a sociomental space, mentally anchored "in" our portable communities. The reality of this space for us is thus reinforced and made more tangible.

We can feel completely "out of touch" with others when our portable communities are not at hand for one reason or another:

> I NEED to have my cell with me at all times. If I don't have it, I feel very isolated and out of touch, even if I'm in class

and I wouldn't be able to pick up the phone. I often wonder how people spent time away from each other in the past before cell phones with no way of knowing immediately if something happened to the other person—how did mothers ever allow children to go on trips? (BookLover1)

I constantly carry a cell phone on me, to feel connected to the outside world. If I do not have one, I feel that I am out of touch . . . (WorkGroup2)

When I am out of touch I constantly worry about whether there is something that I need to be taking care of. (Activist)

. . . or we feel left out . . .

I feel like I need to check [my online and text messages] regularly or I'll be left out. (BookLover1)

or we worry we will be seen as "rude" or "mean" if we are not constantly available:

Since people know I always have my cell phone, if I don't answer it or don't feel like answering it, people can think I'm being rude or that I'm angry at them or that something must be wrong. (MovieBuff)

Maybe there is someone you don't want to be in contact with and they keep calling your cell phone and they tell people you're mean because you will not pick up the phone. That causes trouble. (SocialNetworking10)

People report feeling great anxiety if their phones or wireless devices are not on (or on them) at all times:

Not having that phone on you makes some people feel anxious and tense . . . because I think people go around with the fear of being without their cell phone and missing something they believe to be important. (Blogger1)

I carry my cell phone with me at all times. In fact, I actually get nervous when it's not with me. I have become so dependent on my cell phone that I often find myself wondering how people survived without such a thing. (MobileUser4)

Some go to extreme measures to make sure they are always "in touch":

> If I don't have it, I make sure I call everyone I know from a regular phone or email them to let them know I am out of "touch" and offer them other ways to reach me (boyfriend's cell, etc.). (ParentingForum2)

Some say they feel naked without their portable devices!

> I always carry my phone with me, and I feel very naked without it. I honestly feel slightly panicked when I forget it. (WorkGroup1)

> I carry my cell phone everywhere I go, and when it is not working or I forget it somewhere I feel that I'm naked. . . . I personally know that it's not normal to be so attached to something, but I feel that I need to have my cell phone on every minute just in case someone needs me. (MobileUser4)

It is actually fairly common to claim that one feels "naked" when one's mobile phone is not handy.[31] There is a logic operating here. Portable technology is often kept so close to one's actual body—either on the body or within one's grasp—that it may really be seen as a *part* of us. We are "tethered" to the technology; when it is on us or near us we are truly "not alone."[32] When separated from the technology we may feel somehow not whole, not complete—again, lost.

In its own way, email can be anxiety-producing. People often find the sheer amount of email they receive upsetting and unnerving, reporting high levels of anxiety when they face too much email at a time, or, sometimes, when they do not get enough of it, not as much as they expect. We may feel anxious when others don't respond back to us promptly enough, or when certain others respond too promptly, too consistently! Email is an indicator of social status and thus the amount that we receive (and send) can have strong symbolic importance. In addition, both regular email and spam can easily become all but unmanageable, thus increasing anxiety levels and workloads. As I write this, an average email user spends over 30 percent of a typical workday creating, organizing, reading, and responding to email. And the volume of email is projected to grow approximately 18 percent each year through 2012.[33]

> The negative for me is you have to wade through about 20 emails from the group per day, some long testimonials as to

how great the info is and how it has touched their life just to find the post on a great crock pot recipe! (MobileUser3)

Last month there were over 350 requests for information. (WorkGroup7)

I'm getting around 120 messages per day at my workplace. I would say about 80% of these are not very relevant to my workload. Since everyone has got used to the email, they just started to CC every email to everyone else . . . there are quite a few people who're just annoyed by the email tsunami. (InfoGathering4)

Some people, of course, enjoy getting email, and as it can correlate with social status, they feel sad or anxious when there is *not enough* email:

When I used to check online and I had no new responses I would feel sad and my ego was adversely affected. (MobileUser1)

There is some sort of thrill with sitting down at the computer and having thirty new messages. For some, it freaks them out but I really enjoy it. (WorkGroup3)

If it were 300 instead of 30, WorkGroup3 might feel differently!

Computers and portable devices are status symbols in and of themselves. They range from attractive and sophisticated to basic and plain. There is a also hierarchy of features, hardware, software, applications, and types of phone plans. Keeping up with other technology users and ever-changing products and plans can be anxiety-producing and even induce feelings of hopelessness, as one can never stay ahead of the curve.

Anxiety is also produced as we attempt to master the "rules" and norms of modern technological use. We must decide when to take calls and to text others, and when not to. We show deference to others as we grant them mobile phone access to us—giving them our cell phone number, taking their calls, perhaps even showing the ultimate sign of respect: *not* taking a call or even switching the phone off in their presence! Some research indicates that men are more likely to turn their phone off at certain hours, while women tend to leave theirs on all night long, as they are expected to be reachable at all times (in case their children need them, for example). Women are also more likely to phone others in order to give their physical location. Such behaviors echo long-standing hierarchies regarding gender; as with children, we expect to know where women are at all times, and as they themselves so available, greater social control is levied over them.

We must all fairly constantly make decisions about exactly how available we want to be, and when and where we will permit ourselves to be reached. We must also determine the circumstances under which we should call others—for which purposes? with what frequency? how much is too much, or not enough? Such "decision dilemmas" cause more anxiety because they reflect existing social hierarchies, are difficult to negotiate, and are constantly in flux. When communication and contact are nearly always possible, the limits we place on such contact have great symbolic and practical significance.[34]

There are seemingly endless potential dilemmas and hurdles in managing one's information and technology. Individuals and organizations can spend much time and energy managing spam, spyware, and viruses, not to mention ever-available bits of information and ever-evolving technology.[35] SocialNetworking1 feels that

> [t]oo much information can make one really neurotic . . . I've learned to limit my exposure and not get too many game hints (which makes gaming no fun) or listen to too many pregnancy horror stories.

MessageBoards1 looks at this from children's point of view:

> I think children are affected by the unobstructed access to information. I think it contributes to a sense of overwhelm-ingness (is that a word?) I base it on the face that I see kids going through much more despair than I remember, as if they are confronted with so much information at a time when they are getting ready to go out in to the world that it can be paralyzing.

And for SupportGroup8

> It can be overwhelming at times—sometimes just WAY too much information thus causing me to shut down and avoid the group for a while.

It can be hard to escape portable technologies and the information and stimuli that flow through them, even if we want to. "The uses of wire-less communication technology have made it possible for people to occupy their every potentially idle moment," Castells and his coresearchers point out, "whether by checking email at the bus stop or while waiting for a flight, sending text messages when bored, or conducting clandestine conversations or personal research during meetings."[36] We often use time driving in the car to coordinate activities or interact with others by phone, as SocialNetworking2 does:

> I found myself so busy and far from my friends that late night conversations via cell phones was the only method to keep in touch and catch up.

She also points out, though, that making late-night calls helps her feel safe in the car:

> I often drive five hours alone, that drive can not only become boring but also dangerous out of fear of falling asleep behind the wheel. However, I know I can always put in my earpiece and have a conversation with a distant family member or friend.

It seems that even when engaged in solitary activity, as on a long car ride, we often choose connectedness, whether out of habit or a desire to reduce anxiety and feel safer, catch up with others, coordinate activities (many family activities are coordinated in the car), or simply to pass the time in the company of others rather than alone.[37]

Some people claim that their portable communities *reduce* anxiety overall . . .

> My anxiety and fear were greatly reduced when I found this listserv. The anxiety was replaced with knowledge and hope. (SupportGroup1)

and relieve stress:

> Online groups have helped to serve as a huge stress reliever . . . I play games online because they are good for me in terms of releasing some stress due to schoolwork and classes. (WrestlingFan)

> I usually write blogs to free my mind from stress and to exercise it as well. It feels good and it allows me to think in interesting ways which increase my knowledge and writing ability. (Info-Gathering5)

There are as many ways to respond to portable technologies as there are people who use them. Research that focuses on the ways that people use portable technologies to shift and "soften" time and coordinate activities is much needed, as is research into information and technological overload and anxiety. And as privacy has become a scarce resource in modern technological societies, it will be gravely important to continually monitor and study the impact of portable technologies on our diminishing sense of privacy.

THE IMPACT ON PRIVACY

Private information has never been more easily accessed nor more easily manipulated.[38] Electronic messages can be accidentally misdirected or purposefully obtained by those to whom they were not directed. *Surveillance*—the using of individuals' personal data to influence, manage, or control them—is a real and rising concern, with electronic surveillance becoming more and more commonplace. This has implications for individuals, groups, and societies.

It has become easier than ever for people to discover one another's purportedly private activities and spy or check up on one another:

> I recently visited a site because my fiance's brother is on it and
> we were kind of "checking up" on him and his new girlfriend.
> I also found out that his mother got an account on it just to
> keep tabs on what he is up to. (SocialNetworking5)

People increasingly use portable devices to keep tabs on their partners' whereabouts and activities. Examples of this include a boy who accessed calls his girlfriend received on her wireless device and then contacted male callers to warn them off, a woman who reportedly left her boyfriend because she felt he was making numerous calls to her cell phone to check up on her, and a man arrested for using a mobile phone equipped with a global positioning system to stalk his ex-girlfriend by attaching the phone to the bottom of her car.[39] Global positioning technology that can pinpoint the location of the phone to within a few feet is offered by major mobile phone carriers and can be used to spy on others and deprive them of their privacy.[40]

Parents now use portable technologies to keep a close watch over their children. Although children with mobile phones are generally granted greater independence than those without the devices—and at least initially, they tend to see the phone as a symbol of independence—their whereabouts tend to become almost continually monitored by their caretakers, who provide instructions, schedules, and advice at nearly every turn. Communication researchers Lana Rakow and Vija Navarro call this "remote mothering."[41] Children receive mobile phones as gifts from anxious parents who fear emergencies or want to make sure their children do not fall behind their peers technologically (this has been called the "technological dowry").[42] Some parents track their children via mobile global positioning technology and applications that alert them nearly every move.[43] Adolescents and teens with mobile phones do tend to have later curfews and to contact parents on their phones fairly frequently—nearly as often as they call friends.[44] But in exchange for this they are expected to be available and to accept parents' calls (and watchful eye) at all times, as SocialNetworking10 bemoans:

> This is big for parents because they think you are trying to avoid them when you tell them you accidentally left your phone in the car or it was on silent and you didn't hear it. Especially if you tell them you lost service, they think that you hung up their call or ignored them calling, so they yell at you. It's definitely a lose-lose situation.

For Sherry Turkle, the price children pay in terms of personal autonomy is substantial:

> There used to be a moment in the life of an urban child, usually between the ages of 12 and 14, when there was a first time to navigate the city alone. It was a rite of passage that communicated, "You are on your own and responsible." Tethering via a cell phone buffers this moment; tethered children think differently about themselves. They are not quite alone.[45]

Of course, all of us who spend time tethered to technologies are "not quite alone." But children who grow up in this way may be denied personal exploratory experiences and privacy that may be critical at certain junctures in their development. It seems that most children accept this parental surveillance (in the form of checking in frequently) as an unavoidable part of their existence, though some children have turned their phones back in to their parents in protest![46]

In much the same way, we are all forced to accept a fair amount of institutional surveillance in a technological society. Vast amounts of data, including our own personal, social, psychological, and consumer habits and behavior, are tracked and preserved by both private and law enforcement agencies and kept in "digital dossiers."[47] Whenever we use the internet, information on us, including usage and search patterns, can be easily gathered. Employers, governments, and organizations may access the data at their whim, often legally.

Surveillance of employees by employers has become increasingly commonplace in technological societies. Most companies have the ability to monitor employees' web connections, and as many as half may retain and review email messages. "Employees should know that your employer is looking over your shoulder," says Nancy Flynn, executive director of the ePolicy Institute, an Ohio-based training and consulting firm.[48] Several of the 100,000 members of WorkGroup5's work-oriented discussion forum reported there that they had been fired from their jobs "for sitting on the forum," he told me. "Then, they posted on it about getting fired."

Surveillance by governments is a related, and potentially greater, concern. The United States government has reportedly authorized its National Security Agency to collect records of international phone calls and email communications made within the United States without a warrant, and has also requested listings of users' internet searches from Yahoo, Google, Microsoft, and AOL.[49] Even more intrusively, mobile phones have been used as portable electronic "bugs" for the purposes of eavesdropping on users. Software can now be remotely installed on mobile devices without owners' knowledge, essentially turning the phones into microphones that can transmit a conversation that occurs while in the vicinity of the phone to a remote location, even when the phone is turned off—and the FBI has reportedly already taken advantage of this technology in pursuing information regarding organized crime.[50] Perhaps the most stark example of technological intrusion is the use implantation of microchips under people's skin for monitoring purposes. This, too, is already a reality: a Cincinnati security company has implanted such microchips in employees' arms, and legislators in Wisconsin have felt the need to introduce and pass laws that would ban companies from requiring workers to be implanted with such microchips.[51]

On the other hand, some surveillance acts may be positive. One young teenage boy was able to assist in the capture of a potential molester by taking a discreet camera phone picture of his attacker's license plates and transmitting it to the police. Services now exist to send us text messages reminding us to move our cars or take our medication at a particular time.[52] And because the location of a person sending a text message or mobile phone call can be remotely traced, people stranded or lost in desolate areas have been found and brought to safety. Location-transmission devices that can be integrated into mobile systems and automobiles enhance response times in the case of emergencies and can be seen as "contributing to the general safety of society."[53]

Surveillance is a double-edged sword, impacting civil liberties, privacy, autonomy, and safety in numerous and contradictory ways.[54] My interview subjects were not reticent in describing how being technologically "watched" made them feel. The metaphor of the "leash" was invoked more than once:

> Unfortunately the cell phone feels like a leash. I feel trapped. "Where are you?" "What are you doing?" "When will you be back?" "Why didn't you call me, I would have loved to join you?" (MobileUser3)

> I think of cell phones and beepers as electronic leashes. (OnlineStudent1)

The theme of not being able to escape "even in the bathroom!" surfaced:

> I carry my cell phone all the time, and sometimes I hate it because I can't even go to the bathroom peacefully. You can't hide from no one. (MusicLover1)

It has become so difficult to find a place to hide that the automobile has been likened by Ling to the phone booth, a place to which we must escape in order find privacy in which to conduct our conversations.[55]

Portable technologies can invade—or provide—private time or space. They *colonize* our time and our spaces so that we can now work (or play, if we have that option) nearly anytime and anywhere. "The possibility of perpetual contact," Leopoldina Fortunati notes, "risks shaping time into a container that is potentially always open" so that there is no period of time during which we can not be reached.[56] This constant availability can be easily exploited by those who are demanding or needy:

> Though I do find it nice to be able to get in touch quickly, I also despise the aspect that my coworkers can get in touch with me at any point in the day. . . . I have one of those high-tech cell phone/PDAs and anyone can email my work account at any time and my phone chirps to let me know I have an email. (WorkGroup4)

One can begin to feel that constant productivity and availability—even when "off duty" or enjoying leisure time and activities—is a requirement in modern life.

We have become accustomed to filling up nearly every moment with something to do because with a portable device nearby there is always someone to contact or something to do. Increasingly, we multitask, trying to fit more and more activity (and social interaction) into any given slice of time. In a large-scale study by the Kaiser Foundation, Betsy Israel reports, children from 8 to 18 spent an average of six-and-a-half hours a day using computers, TVs, movies, video games, books, iPods, and mobile phones. They switched rapidly among them and often used several at once. Fortunati states that the result of such constant technological activity is that time in general "is becoming socially perceived as something that must be filled up to the very smallest folds."[57]

In modern life, the lines between work, home, and "third" or leisure spaces are increasingly, perhaps permanently, becoming blurred.[58] Castells and his coauthors note that:

> Perpetual contact means that you can be located whatever the hour of the day it is and wherever you are. This, indeed, has created some problems because the boundaries between private life and working activity have been broken. . . . In practical terms, this non-barrier availability leads to some real tensions between the two spheres of life . . .[59]

and, I might add, to tensions on the psyche. While being always "on duty" can be a plus in the economic marketplace, it can be mentally and emotionally draining, especially for those who learned to partition time and experience more rigidly. Younger people who have grown up with technology constantly at their sides do not seem to make the same distinctions among the "zones" of work, home, and leisure, just as they may see themselves as simultaneously producers and consumers of many forms of online and mobile content. It may seem normal to them to be "on duty" all the time or to allow a variety of technologies and sociomental spaces to interpenetrate all of their physical spaces, all the time.

But in this, as always, there are costs. It is impossible to focus on a single issue in all its complexity when giving continuous partial attention to multiple tasks.[60] Depth is surely sacrificed for breadth. "We insist that our world is increasingly complex, yet we have created a communications culture that has decreased the time available for us to sit and think, uninterrupted," says Turkle.[61] And many of my interview subjects do resist the idea that we can and should be perpetually available to one another:

> I don't like the idea that since I have my cell phone with me at all times that I am "required" by friends and family to answer each call. (SocialNetworking2)

I cannot imagine adopting IM anytime soon. I know what it is, I know the conveniences it could provide. I actively don't want them. I don't want people to assume they can find me, and demand a timely response. (MostlyEmail3)

I would not like nor want to be "available" 24/7. I think that having to be "in touch" constantly robs people of the joy of the moment. I find it rude when I am with someone who must be a slave to their PDA/Blackberry etc. and just leaves me sitting there while s/he answers the message. . . . Email has certainly made my life easier to manage, and voice mail is also helpful, but otherwise, I have no intention of being "plugged in." (SupportGroup6)

I don't like the idea of being in constant reach of people all the time and the fact that people get irritated when you're unavailable or unwilling to participate. (SocialNetworking4)

About half of my interviewees say that they do actively place restrictions on others' ability to reach them, while the other half chooses—or feels obligated—not to do so.

I'm not bothered at all if I am out of touch. I kind of like it. (OnlineStudent3)

I like being out of touch sometimes. My husband cannot tolerate it, unless we are on vacation and I insist. (OnlineStudent1)

But even those who refuse to be constantly available to others must respond to the nearly ever-present modern expectation that *they should be.* This expectation—that we will always pick up the phone or respond to our email—is widespread. My interview subject Activist finds herself using creative means to be "out of touch":

I do carry my blackberry all the time. It can sometimes be overwhelming because normally I wouldn't have access to clients and the office until I was back in the office. Now I can be reached anywhere. I had to start purposely leaving my phone in the car over the weekend so I wouldn't walk by and see messages or emails . . .

Because portable technologies are so fully integrated in our lives and provide such a strong sense of comfort, security, and belonging, we must consciously decide when we are going to put our portable technologies to the side or leave them alone altogether. And even then, we may feel guilt or anxiety for having done so. We face frequent "decision dilemmas" as to how available we will be at any given time.[62]

But surely there are definite advantages to what Hans Geser calls "temporary non-connection."[63] When we are out of touch with others, at least for a little while, we can take a much-needed break from the "work" that social interaction and activities, even leisure activities, generally require. We can take time for reflection and renewal, allow technology-induced emotionality to cool, and think and communicate with greater depth and focus, perhaps seeing some things more clearly as a by-product.

> I think we are too connected anymore. People do not have quiet time to think and allow their minds to wander—allow creativity to happen. (OnlineStudent1)

Geser would agree. "Human existence is enriched by feelings of longing or homesickness, by experiences of anxious insecurity about what others may be doing," he proposes, "by sadness when a loved one leaves and joy when he/she finally comes back."[64] If we never leave one another, we can never miss one another, or experience the joy of being reunited.

Even as we are able to remain in closer contact with children, spouses, and coworkers, we have gained the ability to check up on them, possibly unfairly and continuously, depriving them of private space and time, of the opportunity to just not be found for a little while. "There is an increasing tendency," says sociologist Mimi Sheller, "to slip between private and public modes of interaction, as a result of the new forms of fluid connectivity enabled by mobile communication technologies."[65] Blogs are a good example of the interpenetration of the private and the public, as previously private information can be made public on blogs, even as it is communicated in a quite personal way. Things once kept most private—one's innermost thoughts, feelings, even personal conversations—are now routinely published on the internet for the world to see:[66]

> On [one of the social networking sites] one of the wrestlers connects with his fans. So he is constantly posting blogs and I am constantly responding to them and posting blogs to other fans . . . and with some of these fans you get to form an online bond. . . . The satisfaction of blogs is the ability to connect with

people from various places, not knowing them but sharing a passion together. (WrestlingFan)

But several of the people I interviewed expressed skepticism as to the wisdom of exposing so much of one's private life online:

I don't think I am the blogging type, as I would feel self-conscious about writing my thoughts down for everyone to read. (TVFan2)

For InstantMessaging7, blogs represent an erosion of privacy:

Having your feelings and thoughts so easily scrutinized on a blog is a little odd to me. It is also odd that people can comment on anyone's journal entries. Since when is it okay for people to comment on your diary entry? Blog rings feed into the continual erosion of privacy . . . people's feelings and private matters are so easily accessed.

And trust is a prime concern for some:

[Participating in online groups] has made me vulnerable, yet more protective. Since I cannot really tell (unless it's my friends) who is behind the monitor on the other end, I don't take many chances. (InfoGathering2)

I do not trust [social networking sites], I just think that putting too much personal information in that type of site is not a wise move. Some people expose their entire life on these websites. (WorkGroup2)

Some who share very personal things on their blogs say they do so as a form of "therapy" and that they obtain a strong sense of solidarity and community when others comment and respond to their most personal posts.[67] Some even expose the most intimate details of the lives of *others*—friends, family, loved ones, acquaintances—on their blogs or websites. To what extent should (and do) bloggers obtain permission to publish something about someone else's private, everyday behavior? Barry Wellman has spoken of his surprise in coming across a casual conversation that he once had over dinner mentioned, and informally excerpted, in someone's blog. Afterward, he wondered, "Is it ethical to publish private conversations without the speaker's approval? Or has the nature of networked community become such that just as the public has become personal, the personal has become public?"[68]

This is a privacy concern with considerable ramifications. All kinds of private activities, photos, and conversations can now be casually mentioned, displayed, and in effect "outed" on a blog, electronic mailing list, social networking site, or some portable community somewhere. Eszter Hargittai, a longtime academic blogger and sociologist who studies the internet, tells of how she explicitly obtains permission from others about whom she plans to write. Blogging *about* privacy "in the age of blogging," she writes:

> I certainly know that people in my social circles—friends, family members, colleagues—do wonder what I will and will not blog about from our interactions and sometimes even preface comments by saying "this is not for blogging." I always reassure these people that I never blog information about other people without permission and in general rarely mention any names or other identifying information (except to give credit, but I check in such cases as well).

Hargittai goes on to point out the impact of the anonymity of the blogger—or the lack thereof:

> Since I do not blog anonymously there is more social control over what I decide to make public. After all, everything I say reflects on me in return. Outing information about others that many may find inappropriate will have negative repercussions on me. So even if I had no concerns, whatsoever, about the privacy of people around me—but I do—a solely self-interested approach would still dictate that I keep information about others' lives private in order not to upset people and in turn lose credibility and trust in the future. However, such social control operates much less effectively among those who can hide behind the veil of a pseudonym.[69]

For a blogger to choose to make someone else's private behavior a matter of public record without that person's consent is troubling; to do so while remaining anonymous is even more troubling. When one is not accountable for one's own words, there is a lack of social and privacy control; a lack of reciprocity between writer and subject. It is indeed a trust and credibility issue—not just for the blogger but for all of us, as any one of us might be being "watched" or "outed" online at any time without our knowledge and consent. As Wellman has framed the issue, these technologies require us, as individuals and as a society, to define and redefine whether and to what extent the personal has become public and the public personal.

Portable technologies have become so participatory, so expressive, and so indispensable to so much of what we do, personal and professional, that they seem and feel like private spaces. But, of course, they are not. Even though most blogs are intended for a fairly small audience of family and friends, and are personal in their content, they are not private in their structure. They exist in a public space, and the information that flows along their social networks is publicly accessible. Email, text and instant messages, discussion posts, and internet searches can also be easily retrieved by others. But in the emotional, often intimate moments spent in portable communities, few of us stop to think we may be making a traceable mark. There is a long way to go to educate people fully about this issue.

Because people feel relatively safe online, they may not be consciously aware at all times of the extent to which their behavior is or can be made public:[70]

> [On my favorite site] you don't have to tell who you are or where you live so you feel safe and you feel more comfortable. (Gamer2)

For some, the sheer convenience of the technology, along with the rewards and advantages of being online, trump possible concerns over privacy or surveillance:[71]

> I know that the cell phone or blackberry prevents people from ever being out of touch, but I personally think that the advantages that a cell phone brings in convenience and safety far outweighs what I am giving up in terms of privacy. (SuppportGroup1)

The sense of community and comfort found in online and mobile connecting may be so strong as to induce complacency regarding such critical issues for a technological society to confront as privacy, surveillance, trust, safety, copyright infringement, the security of online transactions, and the ongoing fight for civil liberties and equality.[72]

In the continuity of our connectedness with one another, we gain what Anthony Giddens has called *ontological security*—the "confidence or trust that the natural and social worlds are as they appear to be."[73] We require some constancy and dependability in our social experience; we can not emotionally bear the possibility that life as we know it could be turned permanently upside down at any given moment. In their ubiquity and in the constant availability of their members, portable communities can and do provide this kind of ontological security for us.[74] They are always, dependably, there.

But even as portable technologies bring constancy to our lives, they inspire new and different ways of living and experiencing the world together. Space, time, and social interaction are being reshaped, almost continuously. The roles of parents and children, of employers and employees, of governments and citizens, are being redefined. Privacy and autonomy can no longer be assumed. In the constancy of our availability to one another and the connectivity that results, the public and the private; home, work, and leisure; and creation and consumption are all becoming freely comingled—in ways that will require continued and sustained research for us to ever hope to understand.

7

Harnessing Social Interaction
The Control of Time, Space, and People

One of the charms and frustrations of everyday life is its unpredictability. We can't control it, can't force it to be one thing or another, and we certainly can't predict it with much precision. Durkheim has famously taught that social phenomena will always coerce and constrain the individual, but as Berger and Luckmann, Anthony Giddens, and Erving Goffman (among others) point out, it is individuals that make the everyday choices and take the everyday actions that create the societies that then influence our future choices and actions.[1] We want to take some control over the structure of our social lives, but because we do not know how others will act and how societal norms may change, there will always be limits to what we can control.

Still, we try. We don't want to be asked to make an important business decision on the spot, to be thrust into the middle of a debate in which we're unprepared to take part, to discover that friends or lovers have unexpectedly gone away. For those who seek to control social interaction, at least to some small degree, portable communities can provide a great assist. Online and mobile technologies are tools with which we can shape many more aspects of our social interactions than has previously been possible. Just as Caller ID allows us to pick and choose which phone calls we want to take and which telephone interactions we want to have, portable technologies allow us to manage many of the parameters of social connecting. This has a definite effect on us both online and offline, as interactional life increasingly becomes something we choose to experience in set, configured, controllable ways.

In this chapter we look at how we use portable technologies in an attempt to "harness," or gain some control over, time, space, and other people: making things happen where we want them to happen, when we want them to happen, with whom we want them to happen. We examine how online and mobile technologies help us control where, when, and whether we engage in social interaction. Then, we look at the technology-based strategies for interaction used by my interview subjects to make these decisions regarding their social engagements. Finally, we take a look at the resultant state of predictability and spontaneity in social interaction in the modern technological age.

WHERE, WHEN, AND WHETHER WE INTERACT

As discussed in the prior chapter, the internet and mobile telephony permit nearly constant access to at least some of the others in our communities. If we want synchronous interaction and that sense of being with others "right now," we can arrange to IM, visit a chat room, "live blog," or email, text, or talk "live" on the phone. Many of us arrange things so we know when our friends will be available online, so we can "meet" them in one of these spaces. We can pick up the mobile phone at almost any hour of the day and find someone with whom to talk or text, producing the temporal symmetry (see chapter 2) that is such a powerful agent of social bonding.

If we prefer asynchronous interaction, we have even more options: there are literally millions of discussion boards, blogs, social networking spaces, and websites that can be visited at our leisure. Email and text messaging can be engaged in asynchronously as well. When time management is a more pressing concern than immediacy—or when we prefer the luxury of spending additional time and thought on our responses—using technologies in this way can be a satisfying option. We tend to have a general sense of how frequently a given individual accesses his or her email (according to one estimate, half of all emails are accessed within one hour of delivery). Email allows us to be a little less available than synchronous communication requires, but still retain a sense of general availability.[2] Though we will not necessarily be there at the exact moment someone wants to reach us, we can still get back to them reasonably quickly. This can give us a sense of "quasi-synchronicity" in our interactions with others.[3]

With these tools at hand, the individual can literally choose where and when to interact:

> I like the internet. . . . because people can respond when and how it fits into their schedule. (MessageBoards1)

> I take advantage of email where I can write on my own schedule as opposed to real-time. (WorkGroup1)

> The internet acts like a bit of a "shield" where I can initiate and respond when *I* want to, not at someone else's beck and call. (SupportGroup8; emphasis in original)

This "shield" provides a kind of "temporal distance" that gives the user some control:

> I think I like the temporal distance that email provides. Nobody knows when I'm reading it, and nobody is expecting me to reply by a certain time. I like the temporal control over email that we just don't have when text messaging. (MostlyEmail3)

> If someone calls that I don't want to speak with, I simply do not answer! (WorkGroup6)

This kind of control was named by many interviewees as a major perk, a valued interactional benefit, offered by portable communication technologies.[4] As we bring technologies into our private times and spaces, they allow us to exercise more control over one another than we can have in more sequestered arenas.

We do not always use these technologies to make connections with others. Sometimes, we use them to avoid or postpone connection:

> With IM, I can block talking to anyone I don't want to talk to. (InstantMessaging5)

> I do not always answer and recently have stopped carrying my phone to places. (InstantMessaging2)

> Now I always know who is calling (with caller ID) and I never answer a call if I don't know who it's from (I listen to the voice message and can call back) and often don't answer calls from friends/family is it's not a good time to talk (e.g. if I'm out to dinner and a good friend who lives far away calls, I won't answer because I know she'll want to talk for a while, and I don't currently have time). However, if someone calls twice in a row I will pick up, because that means it could be an urgent problem. (WorkGroup1)

It can depict or confer power to make these kinds of decisions:

> I like to *not* answer the phone. I like to know who is contacting me, I want to hear their message, and I want to choose when to return the call. It's all about knowing more about others than they know about me. And choosing when I make contact with them. (MostlyEmail3)

> I rarely call those people who don't have cell phones because I really don't want to talk to everyone else who lives in their house before getting to them. (Blogger2)

> With people I'm trying to avoid, I call and leave a message when I know they won't be around. (SocialNetworking1)

And these decisions can provoke strong emotions:

> If I don't hear my cell phone ring or choose not to answer it my boyfriend or my parents get angry because they think there's no excuse not to pick it up. I don't like that. . . . There are certain times when people need to be . . . unreachable. (MobileUser2)

Caller ID, cell phone ring tones, and similar inventions permit us to decide with precision whether, when, where, and with whom we wish to connect. We can use multiple email addresses and filters to keep out unwanted messages.[5] Even very small, specialized aspects of social interaction can become harnessed.

In a modern technological society, these moments, these instances of interaction, become factors under one's control instead of out of one's control. For busy people or for those who are simply used to interacting in this way (youth, for example), this is a valuable, even indispensable benefit of online and mobile connecting. And it can be significantly anxiety-reducing. Much of the anxiety that can take place when people confront one another face-to-face can be ameliorated when their exchanges are technologically mediated. As psychologists Yair Amichai-Hamburger and Katelyn McKenna state:

> Many of the situational factors that can foster feelings of anxiety in social situations (e.g., having to respond on the spot, feeling under visual scrutiny) are absent in online interactions. Because participants have more control over how they present themselves and their views online (e.g., being able to edit one's comments before presenting them), they should tend to feel more comfortable and in control of the situation. They should be better able

to and to more often express themselves, to be liked more by their online interaction partners than if they interacted in person, and to develop closer, more intimate relationships through online interaction.[6]

Control over social interaction, then, can promote intimacy, as some of us are more relaxed, comfortable, and able to be ourselves when anonymity and distance combine to give us a sense of control over interpersonal situations.

This anxiety-reducing benefit was noted by my interviewee MusicLover1, who told me that

> when I'm at work and have to ask a client or co-worker a question, or I have a problem with a service and need to talk to someone about it, I prefer to contact them through email. I think it has to do with the fact that I'm not sure of myself or that I'll know the right thing to say. I'm one of those people who, once a discussion is over, thinks of a million things I should have said. When I email someone, I know that the email I send will communicate exactly my intent (maybe not my emotional intent, but at least my factual intent).

The more comfortable people are with online and mobile technologies, the more they look to those technologies for a sense of control over situations.

Although controlled (and controlling) social interaction can be highly satisfying, the evidence does not suggest that we as a society are in danger of technologically mediated interaction supplanting face-to-face association. As discussed in chapter 4, though people often initially become somewhat immersed in portable technology use, such technologies tend to become integrated over time into the rest of one's life and are often used to facilitate face-to-face meetings and dates. Doing this requires us to develop specific strategies for determining exactly how we will use each portable technology—separately, together, and in combination with face-to-face interaction.

TECHNOLOGY-BASED STRATEGIES FOR INTERACTION

Members of portable communities find that they can establish a means of controlling the technologies and applications they use in social interaction and of allocating them to the various interactional tasks they face. As we become more adept at using them, we learn to assess which technologies we want to use for different types of messages, using the set of them as a sort of menu from which to choose. As internet and mobile users we must parcel out what we have to do and say across the range of technologies we use and throughout our time spent offline as well. We do all this in the manner that reflects our needs and wishes and works with our lifestyles.

Though we may not be aware of it, we each develop and execute a number of often intricate technology-based strategies for interaction. This gives us a feeling of control over the many, many exchanges that are made in the course of an often overfilled day. These strategies are not always the same from person to person, and vary almost as widely as the people that use them. But without them, professional and personal lives would be almost impossible to manage.

Most of us, for example, develop and utilize strategies for when we will and will not use telephones (landline or mobile). We try as best as we can to control who has our phone number, and may even assign different people different numbers (family may be given one's home number, while colleagues may be given the cell number—or vice versa). A hierarchy can easily develop in which those who are closest are given certain numbers (or, for that matter, certain email addresses). This permits us to limit our accessibility to some people while we grant it to others.[7] Some people prefer to save their most personal interactions for the phone:

> I'll email friends with brief messages to see how they are, but I usually prefer speaking with them on the phone to really "catch up." (MusicLover1)

> My preferred method of contact is the telephone with people that I know well as I can do it while I am driving the car. (MobileUser1)

> When I like to be more personal, and have a real conversation, and I'm in my free minute zone, I will use the phone. (Lurker2)

> Using a cell phone is something I do every day because it is one of the only ways of communication with my family, friends, and even boyfriend . . . (Blogger1)

> I use the phone the most and prefer to talk than to type. I've noticed that emails, across the board, are often misunderstood . . . (SocialNetworking4)

Some prefer text messaging:

> I like text messaging as opposed to phone conversations because there's no pressure to have a whole conversation and I can be pithy. (SocialNetworking1)

> Text messaging is an addiction I'll admit I have. I am not a big phone talker, but I do like to communicate with friends similarly, so I'll go back and forth with text messages. Texting and IMing give you an easy way to connect with people. You won't bother your ears by having the phone attached to it all day long, and it's cheaper than calling. (WorkGroup4)

For many, texting is similar to instant messaging:

> IMing allows a slower dialogue. The slower speed can be beneficial—more thought goes into the conversation. (Online-Student1)

> Using IM to catch up with someone is much cheaper and easier than speaking on the phone. [And] it lets you have a minute pause between your next sentence. A minute pause on the telephone would be just awkward. (InstantMessaging7)

In text and instant messaging, there is sufficient time for editing and self-reflection before responding to a message, yet we can still expect that our messages will be received and responded to within a reasonably short period. We also know that we will not be interrupted in most textual communication. Email, IM, and text messages often play a major role in helping us remain connected with our face-to-face loved ones. Groups of connected individuals sometimes text one another sequentially in semi-organized text circles, using "reply to all" to keep everyone in the loop.[8] My mom, brother, sister, and I, along with our spouses, do this frequently, as an efficient way to share mundane or important (or just funny) observations or information with everyone in our little community all at once. Whenever another circle of emails begins to go round and round, we are brought more tangibly into one another's thoughts, our connectedness over many miles strengthened.

Some people prefer to put their most personal thoughts into writing. They may feel more comfortable with the written word, believing it gives them greater distance from their subject or greater anonymity. Or they may prefer to interact without time pressures or interruption.[9] Using email, people can plan what they want to say:

Email allows me the flexibility to write, which I sometimes like better than speaking since I can think out each statement before writing it (instead of uttering words I wish I hadn't of said)☺. (MostlyEmail1)

Most people are too in love with the sound of their own voice anymore, and if I'm talking to them and stumble over a word or am getting my point out too slowly, they will take that opportunity to interrupt me and start talking. In email, I will get my whole point out. (TVFan1)

WorkGroup2 appreciates email's permanence:

If I want to be able to verify or be able to refer back to what I discussed in a conversation, the best method would be through email, in order for me to print out the conversation or refer back to it at a later date.

Email remains the most popular web-based application, although it is becoming passe among younger users, who prefer instant and text messaging, especially when communicating with peers.[9]

Some people write blogs to serve essentially as mass emails, to efficiently keep family and friends, their primary intended readership, up to date on their latest activities and musings:

I write my own blog . . . it often is just an update of what's going on in my life. It's so my friends from out of state or out of the area can feel like they are keeping in touch. (BookLover2)

If there is something you want to get off your chest . . . you could tell your closest friends over the internet in a blog and they could comment back to you. (WorkGroup4)

Most blogs are targeted to a narrow readership:

Blogs are certainly the most egotistical exercise in online interaction. . . . I started mine because the woman I was seeing met someone on the site and so I know they were both on it regularly and might read it. It allowed me to vent my feelings about them, women in general, dating, etc. without having to be direct or confrontational. (SocialNetworking1)

> Blogs are usually about what you did that day, relationships, friends, dreams and even future goals people may have. A popular thing is to also write about who you like and who you don't like. . . . to vent out my feelings at some points and it was a way to let others know how I felt without having to tell them directly. (Blogger1)

Bloggers can now create an interactive, multimedia kind of correspondence, complete with photos and even video, that is permanent, archivable, and easily linked to other sites. This is a strategy for communication and self-expression proving very popular.[10] Blogs like this "function as asynchronous emails to people who are already emotionally involved in the blogger's life," observes blogging pioneer Rebecca Blood.[11]

For those who prefer to take even more time and care in writing, old-fashioned letters are not yet extinct:

> Letters are reserved for the most intimate of friends and family, or used as proper responses to gifts . . . (SocialNetworking1)

> If I want a more personal communication then I will send a letter. (WorkGroup2)

> People don't expect them anymore and they are nice to get! (MessageBoards2)

SocialNetworking4 provides an excellent example of how we exert control over even the letter-writing process:

> I refuse to give my cousin my email address because I love to get her letters and send letters to her (she lives in Europe).

Some people prefer the highly contemplative nature of writing to one another in letters or lengthy emails. This gives them the space and time they feel they need to share personal thoughts and feelings appropriately and accurately. Letters, in particular, seem to imply great thoughtfulness on the part of the writer. In fact, most of my interviewees who still write letters do so to keep in touch with their grandparents![12]

Sometimes we use several technologies at once, or one technological application with several people at the same time. This is common among younger users of media and communication technologies. But connectors of all ages are prone to multitasking in this way, attempting to pack more and more into a given sector of time.[13] I encountered many instances of this among my interview subjects (and their family members):

> My younger sister (she's 14) is more likely to spend hours instant-messaging her friends than talking to them on the phone. She says it is better because she can talk to more people at once. (SocialNetworking7)

> I often see children multitasking, using several types of electronic devices at once. . . . unlike my kids, I find it difficult to carry on more than one IM conversation at once! (OnlineStudent1)

The necessity of multitasking, at least at times, is not lost on most of us . . .

> I use the internet while I am at work or multitasking at home . . . (WorkGroup1)

> I text message constantly. It just fits into my schedule better than making several phone calls. I can have several "conversations" at once using text messaging while still continuing my normal activities. (MobileUser6)

> I can multitask, do my work, and at the same time talk to a few people at one time (as long as you don't confuse your windows!) and do other tasks at the same time. (WorkGroup6)

. . . as it helps us to remain in touch with one another . . .

> IM provides a medium to carry on several different conversations with several different people at the same time, which can be both a drawback and a satisfaction. It can be a drawback if you get spread too thin and would prefer to have just one in-depth conversation at that time; it can be a satisfaction because you feel connected to more people. (SocialNetworking7)

who may even be "shamed" into multitasking, feeling that there is something wrong with us if we aren't particularly good at it:

> Believe it or not when I first became a part of the online community I couldn't talk on the phone and have an online conversation at the same time. I have mastered it now. (Chatrooms1)

The challenge of multitasking online is to keep up with, even "master," all one's simultaneous conversations. Usually, if one of the conversations becomes too intense, multitasking stops.[14]

When deciding which type(s) or application(s) of portable technology to use at a given time, we consider and then prioritize several factors:

circumstance:

> I use whichever seems most appropriate for the situation. (SupportGroup4)

convenience, or ease of use:

> I use email for work or to avoid calling someone I don't particularly want to have a long chat with. The phone is reserved for immediate matters and chats with folks I'm not trying to avoid. (SocialNetworking1)

> I use the cell phone to call someone for directions, or write an email to get a short message across that wouldn't require a call. (SocialNetworking8)

> I usually use the cell phone . . . It is the easiest way to get ahold of me and I don't have to worry about getting messages. My house phone messages are often deleted by my ex-husband or my children forget to tell me if someone has called. (OnlineStudent4)

speed of use:

> I usually use email for all school-related communication; it actually seems to get answered quicker than phone calls, believe it or not! (Blogger3)

> Usually I use the phone and internet because they are the fastest method of getting in touch with someone. The communication is split second! (ParentingForum2)

> Internet communication is the way to go. I barely ever use the phone because it is so easy to shoot an IM or an email to someone. (InfoGathering6)

> Usually, if I am on the computer and I need to talk to someone about something, I check to see if they are on the computer as well. If they are, I think that I tend to chat with them on the internet quickly, rather than deciding on a time to meet up, or spend more time on the phone. (InstantMessaging1)

monetary expense:

> My listserv and chat room allow me to communicate without cost with anyone, anytime. (SupportGroup2)

> Face-to-face interaction is planned out now because there's so little time on weekends, though gas prices are starting to affect that too when someone lives further. (MobileUser5)

time of day:

> I use email a lot because it allows me to communicate at a time convenient to me and others. For instance, there are times when I would want to communicate with my sister but know she is having dinner with her family. Using email, I can communicate with her. (OnlineStudent1)

> After 9:00 I am constantly using my phone . . . I call friends and talk to them for a while (since minutes are free). (Info-Gathering3)

and physical place:

> The phone gets used at home, cell at school or in a store. (MobileUser3)

> I use a regular telephone in an office, I use my cell phone for all contacts outside of work, I use instant messenger quite often when I am home. (InstantMessaging7)

These variables, and others that should be uncovered via additional research, all factor into our technology-based strategies for interaction yet are employed differently by each of us according to our needs, circumstances, personalities, and social situations.

We also make decisions based upon those with whom we will be interacting. Depending on who is at the other end of a phone, computer, or PDA, we adjust our strategies accordingly. We consider . . .

the nature of the audience at hand:

> I tend to look at my audience I am communicating with ... if the receiver gets a lot of paper letters on his/her desk, and I want my letter to stand out I may send it thru email. (Social-Networking2)

> If it is business-related with a potential client, I would prefer a face-to-face or at the very least a phone call ... (Online-Instructor)

> Emails can be used for business, unlike texting. (Social-Networking1)

> People who do not have computers, I call. People with computers I most often write email and may call every once in a while. (SupportGroup1)

> My main form of communication with personal friends and family is cell phone, and with clients it is a combination of cell phone and email. In fact now with my blackberry, I use email and cell phone a lot since they are so nicely combined in one. (Activist)

and the type of relationship we have with them:

> It depends on the level of closeness with the person ... If it's my best friend's birthday, of course I will go see her face-to-face, but if it's someone I haven't seen in a long time and I'd feel awkward calling out of the blue, I might send a text message or email. (BookLover1)

> The relationship I have with the person determines what type of channel I use. (SocialNetworking6)

We assess other people's different capacities and interests in using portable technologies and our relationships with them, and adjust our strategies accordingly.

We also consider the message we will be sending at any given time and the most effective means for sending it. We look at ...

the nature of the message:

It all depends on what needs to be said. (MusicLover3)

I chit chat on the phone, send documents via email, and use snail mail for things that cannot be emailed or discussed on the phone. (MobileUser1)

I use email to send long stories or an "I miss you" email. . . . I send messages on [a social networking site] privately to quickly tell someone something that doesn't need to be responded to. (MobileUser2)

If a more serious issue is at hand, I tend to use the phone or personal, face-to-face meetings. (InstantMessaging1)

the formality of the message:

The internet is for casual conversations. . . . I text the people closest to me inane texts about my day. The typical "how are you doing, how are things in your life." (InstantMessaging5)

I use the phone or face-to-face to communicate more intimate details. However, if it is a sense of detachment that I'm seeking, I use email. (MostlyEmail1)

and the importance of the message:

I use the internet if I don't really NEED a response but just want to chat. (MovieBuff)

The phone is for some important information, and sometimes just to say hi. In-person is to spend time doing something with someone, or if it is something serious. (InstantMessaging5)

I mostly use face-to-face and telephone communication for important things. I use internet communication mostly for entertainment purposes. (Lurker3)

Blogger2 uses face-to-face interaction only for very particular purposes:

A face-to-face meeting is appropriate only when there are fun plans involved, or when a serious discussion which may include physical contact is necessary (such as breaking bad news or good news, and then you'll want to hug each other).

. . . while others prefer to reserve it for important people and events:

> i base my communication medium on how important the topic
> is to me. if something is really important, i will wait to have
> that face-to-face confrontation. (SocialNetworking5)

> Face-to-face is for the closest of friends because they are the
> ones I want to spend physical time with. (SocialNetworking1)

Face-to-face interaction remains, if not everyone's preferred form of interaction for every purpose, still a special, meaningful form of interaction.

Though strategies for interaction clearly differ from person to person, nearly everyone has some kind of strategy. We all go through a very individualized process of reasoning when deciding whether to email others, text or instant message them, call them, see them in person, or write a letter or note, etc. We make countless such decisions everyday, at work and at home, though we are not always aware of it:

> I'm sure there is some logic to it, but I can't think of it now.
> I email people, but of course I also call them and we meet, so
> no idea. (ParentingForum2)

In the end, it is a fascinating area in which research is much needed: how and why we decide upon forms of interaction and how technology is used to help us make (or even to "drive") these decisions. For it is clear that technologies are used in different ways and different combinations by each of us as we design our strategies for interactional life.

These strategies then become incorporated into our portfolios of social connections as we decide just how, when, and where we will interact with everyone with whom we are connected.[15] For just as we must keep our busy schedules straight, so must we control and organize a packed portfolio of social connections and communities. It might seem that spontaneity would suffer within such a system. But, interestingly, portable technologies can actually return a sense of spontaneity to our often overscheduled lives.

SPONTANEITY AND SOCIAL INTERACTION

Modern lives are notoriously busy, densely packed with all sorts of events and obligations and people coming and going. Many of us are almost habitually busy, for a host of reasons both logistical (we work long hours, have numerous other obligations) and sociopsychological (it is a kind of high, even a status symbol, we're afraid we'll be left out if we slow down, we're avoiding other issues, or we may simply not know how not to be busy). And a prime reason that we become constantly occupied is that with portable technology so readily available, always by our sides, it's so easy to fill every moment with something to do.[16] Technology is both a contributing factor of and response to this kind of hyper-scheduled lifestyle. As we have seen, social interaction is highly controlled and configurable. At the same time, technology facilitates microcoordination, permitting us to change and rearrange plans at the last minute, even as an event is in progress, changing our minds over and over again. What, then, is the state of spontaneity in the face of all this control?

Spontaneity is most often a by-product of flexible and informal, rather than fixed and formal, social interaction.[17] Because portable technologies are so often on our bodies or by our sides, they inspire this kind of use: flexible, frequent, informal. Mobile phones in particular make possible a more fluid, spontaneous lifestyle, as Fortunati notes regarding their use in Italy:

> It is a particular feature of the Italian sociality that one must neither be nor appear programmed. Regimented living and precise organizationally planned activities are abhorrent. Rather one must be and must appear to be engaged in spontaneous activities with a posture of openness. . . . This inherent sense of . . . flexibility, which can also appear to outsiders as disorganization and incoherence, leads the mobile to be seen as the ideal instrument for rapidly adjusting the organizational fabric of daily living.[18]

Because they permit microcoordination and flexibility in scheduling, portable devices can return spontaneity to much of our lives (indeed, even to our culture at large).[19] The use of asynchronous communication, instant-messaging, and humor have all been associated with online spontaneity as well.[20]

Some of the spontaneity inherent in social interaction (especially in informal settings) also counteracts some of the predictability that we can impose on life in portable communities. Though we certainly control much of the *structure* of interaction in portable communities, interaction by its very *nature* contains elements of uncertainty and spontaneity that are impossible (and would ultimately be undesirable) to quash.[21] When people get together, anything can happen. Social relationships retain, at their essence, a defiant element of spontaneity that is impossible to fully strip away.

Relationships are also rather wily. They contain a certain amount of deception that is inescapable, nearly universal.[22] We can not share our every thought even with those closest to us. We are constantly making decisions as to with whom we should share various pieces of information. A certain level of secrecy is a part of social relationship,[23] as is deceit.[24] The extent to which one can predict another person's behavior or response is therefore necessarily reduced; we can never be completely certain whether a person is withholding something that they consider private, or even behaving with outright deception.

But that is the nature of social life. In our interactions with others we will be deceived sometimes. We will be surprised, hurt, disappointed, misled. We cannot control much of the content of social interaction, which is perhaps why we look so readily to control some of its structure. It is certainly possible that over time we as a society may become increasingly less tolerant of imperfect, unpredictable, uncontrollable face-to-face interaction. We may try to control more and more aspects of more situations. We may look to use portable technology to harness interaction as tightly as possible. Conversely, we may choose to use the technology to create more moments in which we can feel more spontaneous, more free, than ever before.

As portable technologies are so easily transported and brought with us anywhere, we now reach out to one another much more often, if briefly and sometimes superficially. Still, this frequent use—this ability to call someone spontaneously wherever we are, to text someone at any time to say "Good morning," "I'm thinking of you right now," or "Good night"—encourages spontaneous use and possibly even spontaneity in other aspects of life as well. People are quite possibly in greater contact with one another than they have been since the advent of industrialization. Though we are often physically separated from one another, we can now quickly and spontaneously make contact, even if we have nothing particular to say and just want to feel one another's presence. As Kate Fox argues, this kind of quick, frequent talk

> restores our sense of connection and community, and provides an antidote to the pressures and alienation of modern life. The space-age technology of mobile phones has allowed us to return to the more natural and humane communication patterns of pre-industrial society, when we lived in small, stable communities, and enjoyed frequent "grooming talk" with a tightly integrated social network. Mobiles are a "social lifeline" in a fragmented and isolating world.[25]

In this view, technology is helping us reestablish the kinds of communal relations that were once much more prominent in our society.

Emile Durkheim called for a "new organic solidarity" in modernity that would mediate between the individual and the state and ensure societal cohesion and integration. Portable communities may be perfectly placed to contribute to this form of organic solidarity. As they are not the province of any one state or physical region, they are situated to help highly differentiated, urbanized societies to structurally cohere and their members to become more fully integrated into them.[26] The technological advances represented by portable technologies, Fox says, can "counteract the adverse effects of previous technological advances."[27] They can encourage the frequent, spontaneous forms of social connection that can tie us together as communities and as a society.

The lives of portable technology users are in many ways both physically and sociologically nomadic. We are always on the move. With portable technology at our side we can bring an element of control to the many interactions we must make and to our lives in general. We have much to learn about how we harness social interaction in portable communities, and the effects on "the rest of our lives" when we do so. As we become more educated as to the internal and external dynamics of portable communities—and educate others in turn—we should work to understand the effect of these dynamics on our own ability to manage our overstuffed portfolios of social connections. Ideally, we will use this knowledge to help us calibrate lives that contain the proportion of spontaneity and predictability that is most workable and functional for us and for our societies.

8

CREATING, EXPRESSING, AND EXTENDING THE SELF

(and Watching Others Do So)

Human beings have a strong need for self-expression. We are almost continually involved in what sociologist Erving Goffman called *impression management*.[1] We act in ways that we believe will convey certain perceptions of us, in the hope that others will see us as we want them to see us. Portable technologies are frequently enlisted in this effort to influence others' perceptions of us. We use them to create personal expressions of all kinds and then to carefully manage these impressions as we communicate them to the wider world.

We are also involved in a lifelong process of identity and self construction. *Identity* refers to one's internal self-definition with regard to our traits, values, beliefs, and the social roles we play.[2] It is also sometimes seen as a performance, as acting as and then being recognized as a "certain kind of person"—which is an external, not an internal state.[3] The *self* is one's individualized conception of one's own personhood. We build a self in and through our interactions with others.[4] Both the internal development and external presentation of our selves (and in a larger sense, our identities) takes place in communities. In turn, "care of one's own identity, one's reputation, is fundamental to the formation of community," social media researcher Judith Donath points out.[5]

In the modern age, self-development and self-expression is often a kind of creative project. Technology is called upon to help us create and express our selves and then to extend and share these selves with others. Portable technologies turn out to be so well suited to these tasks that there has been an explosion of self-development and self-expression in online and mobile use. And we not only seek the expression of our selves in portable communities, we watch (and often become quite emotionally

159

involved) as others do the same. We are both identity *creators* and *consumers*. We then become even more intertwined with others, even more tightly linked together.

Portable technologies give us the means to make and remake our selves and even play around with highly personal aspects of the self such as gender, race, sexuality, and age. And they make it easy to "voyeuristically" peek in on others expressing their selves as well. This chapter examines the ways that modern individuals use portable technologies to construct and express their selves and identities and construct social connections and communities that can be especially personal and powerful. We will trace the processes by which portable technologies enable socialization, the making and remaking of our selves, voyeurism, and the hyperlinking of the modern identity. In short, we take a closer look at how are selves and identities are made and expressed, viewed and shared, in portable communities.

SOCIALIZATION IN A TECHNOLOGICAL AGE

Socialization is the lifelong process of becoming simultaneously who we are and a member of society.[6] These are complementary processes. It is in social interaction with others that we discover and develop the qualities unique to us, and in which these bloom and grow.

Other people, then, are critical in the development of a self. Through interacting and identifying with networks and groups and communities of others, we incorporate some of what we see and experience into our own identities. We observe others—their attitudes and behaviors—and incorporate some of what we see into our own identities. We take on the roles of others—"trying them on for size" much as we may try on our father's shoes or our mother's makeup—in preparation for finding our own way in the world.[7] We also use others as a mirror, adapting our own attitudes and behaviors in response to the reactions that we perceive others are having to us.[8] On and offline, we meet others who become the role models, mirrors, and sounding boards that we use in the slow, gradual shaping of our selves.

Socialization begins at birth (and even beforehand), as we are prepared for the society we will enter by those who bring us into it. Preconceived notions of gender, sexuality, culture, and society are subtly inculcated by those who teach us, play with us, dress us, and feed us. We dress female babies in pink or ruffles to symbolize (and bring into being) their "femaleness," and we make equally sure not to dress male babies in ruffly pink dresses, to underscore their "maleness." Information about what it means to be a girl or a boy, a woman or a man, or any of a thousand social characteristics continues to be subtly (and often not so subtly) communicated throughout our lives. The social climate into which we are born thus influences our identities.

Many children are now born into a world infused with technology. In a host of ways, these technologies color the identities of those who use them. We are now exposed to many others who can influence us. We get to know some of these people directly; others we merely learn about. We can then take on their roles from a distance, even if we have never met them physically. And groups of people whom we are exposed to via mass media and computer and mobile technologies can serve as reference groups from which a host of potential roles and traits can be extracted.[9]

Children's experiences with technology can have a tremendous impact on their identity.[10] When one has grown up with access to technology and develops the skills to use it, it becomes a part of the child's environment and can influence him or her dramatically.[11] Gamer1 describes this well:

> Being involved in online communities was a great time sink in my adolescent life, and it still is. Also, while intellectually I've matured beyond my years, emotionally I'm still just a teenager. I don't know if online communities helped or hindered this. I do think they helped me find my sense of self, however, and allowed me to craft an identity I liked within the "safety" of the internet.

Technology is a kind of window on the world—permitting us to know of and encounter many more people than we otherwise might, and our identities to be correspondingly explored and expanded. Children, in particular, tend to look through this window quite a bit.

Technologies of portability are a major part of many children's everyday lives. Children often receive mobile phones at very young ages from parents who do not want the children to be on the wrong side of a "digital divide."[12] They use portable devices to remain in near-continuous ambient copresence with their friends, make fashion statements,[13] and even adorn their bodies.[14] Children, adolescents, and teens also enjoy expressing their creativity online: at latest count, fully half of *all* American teens had created blogs or web pages or made or shared original content online in the form of artwork, stories, videos, or photos.[15]

Online and mobile technologies serve as an important source of social interaction,[16] health information,[17] learning,[18] and identity exploration[19] for children. In addition, a significant number of adolescents have viewed sexually explicit material on the internet and on mobile phones.[20] Specialist in children's online experiences Angela Thomas reaffirms the importance of identity construction and socialization in early years and the effect of technology on this. "Young people are engaged in struggles of identity formation: they struggle for power, popularity, to define who they are, and to understand their sexuality," she notes. "This is reflected in their online worlds."[21]

The process of identity construction gains speed and urgency in the pre-teen and teenage years. As danah boyd, who has studied the ways that younger people interact on social networking sites, describes it:

> Youth look to older teens and the media to get cues about what to wear, how to act, and what's cool. Most teens are concerned with resolving how they perceive themselves with how they are perceived. To learn this requires trying out different performances, receiving feedback from peers and figuring out how to modify fashion, body posture and language to better give off the intended impression. These practices are critical to socialization, particularly for youth beginning to engage with the broader social world. Because the teenage years are a liminal period between childhood and adulthood, teens are often waffling between those identities, misbehaving like kids while trying to show their maturity in order to gain rights.[22]

Young people face numerous identity issues. Portable peer communities can be vehicles of self-discovery for them. In them, they receive peer validation, feedback, and friendship. These spaces also serve as hangouts where children interact with their friends. There are relatively few physical places where adolescents and teens can freely go, especially when they want to escape parental authority and supervision (an important factor in the development of maturity and a sexual identity).[23] According to Sherry Turkle,

> At the beginning of adolescence, young people have always been confronted with an overwhelming set of new circumstances. There's a new body, new social demands, new peer pressures, new relationships to forge with parents. At this point, with all of this going on, young people have traditionally sought a safe place where they could have a sense of total mastery and a place where they could experiment in relatively consequence-free ways. Our culture presently offers precious few such spaces.[24]

Part of the reason that internet and mobile phones are so popular with youth is that they provide them with these kinds of spaces. In addition, with portable devices in their pockets they can move about more freely in physical space and feel more independent. Many parents feel more comfortable allowing children to go out when they are in touch by phone, although as I have discussed, children sometimes feel unreasonably tethered and constantly checked up upon as a result.[25]

Adolescents and teens act in portable communities in ways that suit their preferences and lifestyles. Younger computer users tell us that instant messaging and text messaging serve their desire for faster and more continuous contact and for maintaining their friendships and social networks.[26] They often "mix" media: using all kinds of technologies together, sometimes simultaneously, to connect with others.[27] Just as in the days preceding modern technologies, kids like to be in touch with their friends—the more, the oftener, the better. They even like to incorporate technology into their face-to-face contacts with one another, sharing text messages and camera photos with one another during their face-to-face "hangout time." This increases the level of interaction and cohesion of their groups. "Connecting with others is the essence," says Rich Ling of teens' use of mobile phones. "Where harried adults may try to reduce the interaction with the outer world, teens try to increase the interaction." Doing so shows that they are popular and in demand.[28]

Some social networking sites (currently MySpace and Facebook) are particularly popular among younger people. They may check their profiles and home pages several times a day, just as many adults check their email repeatedly during a typical day.[29] They like to make and share their own media creations, keep their own blogs, and visit friends' blogs in return. They tend to find blogs easy and fun to make and maintain and easily responded to by peers. In addition, blogs provide links to other blogs, all of which "can foster a sense of peer group relationships."[30] For many teens, using portable technologies socially is a large part of their everyday lives. They spend time in portable communities "because their friends are there and they are there to hang out with those friends," boyd says—which is, again, not so different from adults' motivations for online and mobile connecting (see chapter 4).[31]

Portable devices tend to have especially strong symbolic meaning for youth. Young people are particularly fond of customizing technology to express themselves. They create avatars, buddy icons, customized screen savers. They alter their phones' appearance and ring tones.[32] They customize the language they use in portable communities as well, adopting writing shorthands and creating abbreviations and acronyms that suit their desire for quick and direct connectedness but that also differentiate them from adults and from anyone who does not share their symbol system.[33] Though these linguistic or symbol systems may seem strange, the creation of distinct languages and symbols is actually how all communities differentiate themselves from one another. Although overuse or misuse of these shorthand online languages can have an impact on more formal writing in school or other more formal settings . . .

> Unfortunately, sometimes I catch myself using this short-ened communication in some of my most important school work. . . . For example the word "are" becomes "r." . . . It starts to become a habit that is hard to break. (MobileUser1)

. . . at least one study has found that children that text more often score better than their counterparts on reading, writing, and even spelling tests. It has been proposed that the use of abbreviations might actually be teaching children language skills in an accessible, enjoyable way.[34]

These symbols serve as "identity clues" and are varied and ever-present in portable communities.[35] Many of the people I interviewed were concerned about the impact of online and mobile technologies on their children's socialization:

> Their peers are more "connected" so children need to use the internet to stay connected socially, whereas many adults can abstain from internet use. (BookLover1)

> I think children benefit more [with technology available]. Kids are so ready and anxious to learn about everything that when presented with a fun way to learn they jump on it. (Support-Group8)

> I strongly believe the fact that that children are growing up with computers in the home is shaping the way they view life. I did not grow up with a computer. Therefore, I had to entertain myself in other ways. (MusicLover1)

Children who have grown up with online and mobile technologies are generally quite comfortable using them. A technological world is the only world they know.[36]

Children are using and becoming comfortable with computers, wireless technology, and mobile phones at ever-younger ages, often teaching their parents and caregivers how to use them and becoming the "tech experts" in the family.[37] Many parents worry about keeping their children safe online or about overuse; they seek and receive expert strategies to do so ("no computer in the bedroom," "use filtering software"), and may feel more comfortable if their children carry mobile phones with them wherever they go.[38] Of course, during their life spans, children's patterns of technological use will change, as MobileUser1 discovered:

> My two daughters were "addicted" to chat groups and IMs to their friends from ages 13–15. Once they could drive a car, they were rarely home, so they would choose to speak on their cell phones rather than email . . .

There are undoubtedly effects of using online and mobile technologies at earlier and earlier ages that may be neither immediately seen nor easily predicted. While the desire to be in near-constant contact with their friends may to some extent dissipate, technology use is likely to remain fairly heavy for this population, as it is normative for them to use portable technologies for so many purposes.[39]

Communication technologies have always influenced children's socialization. As SocialNetworking1 says,

> Young ones may be more adroit at computing having grown up with the internet, but they, no doubt, use technology for the same reasons as we do, and probably with the same effects.

And according to WorkGroup1,

> I am only 24, and yet it seems slightly younger people are much more likely to be members of social networking sites, and they seem to IM much more than I do . . . I think I just missed the boat!

Though favored technologies and applications will change over time, they will continue to surround children and teens, much as less portable technologies like landline phones, stereos, TV, and movies have influenced those of us who are older. Our environments have long been drenched in media and technology. Children—and adults—will use the latest, most accessible, "coolest" portable technologies to express themselves and assist them in making friends. But face-to-face interaction remains important to children as well.[40] They may even "use what they have learned online to make better relationships offline."[41] To study the extent to which this may occur and to trace the impact of ever-emerging technologies on young people's patterns of identity construction are but two areas related to technologized socialization in which research will always be needed.

THE MAKING AND REMAKING OF THE SELF

As socialization continues throughout the life span, our selves become created, expressed, and then re-created (and re-expressed!). We create who we are and will be—and yet remain, somehow, "ourselves" throughout.[42] Online and mobile technologies have become prime facilitators of the construction of identity. People are able to express themselves so freely and creatively in portable communities because applications that support and encourage personal expression (and the sharing of such expressions) are ubiquitous and increasingly accessible, cheap or free, and relatively easy to use. In addition, as noted earlier, portable communities are fairly disinhibiting environments.[43] The anonymity or partial anonymity that can be found online—the lack of physicality, the ability to more freely "play"—seems to jump-start self-expression as well. In fact, one of the most commonly cited reasons for initiating relationships online is that "I can express myself better over the internet."[44] This was borne out by my interviewees, who told me such things as

> Personality definitely comes through on the online forums. (MessageBoards1)

and

> I have had some deep conversations on IM that I don't think would have happened in person because you aren't as vulnerable. (Lurker3)

As we allow ourselves to become more vulnerable, more playful, more sociable in portable communities, we can begin to express ourselves in very personal ways.

Blogs are becoming increasingly popular vehicles for self-expression (as well as for more instrumental purposes) among people of all ages. Blogs "provide an unexpectedly intimate view of what it is to be a particular individual in a particular place at a particular time . . . nothing less than an outbreak of self-expression on a worldwide scale (that) empowers individuals on many levels," declares Rebecca Blood, who has interviewed many bloggers on the experience of blogging.[45] "Read any weblog for a few weeks and it is impossible not to feel that you know its writer."[46] Or as another prominent early blogger, Cameron Barrett, shares:

> You see, CamWorld is about me. It's about who I am, what I know, and what I think. And it's about my place in the New

Media society. CamWorld is a peek into the subconsiousness that makes me tick . . . an experiment in self-expression.[47]

Though there are now millions of blogs accessible via computer and mobile device, covering every conceivable topic, they continue to serve primarily as a mode of creative, personal expression: documenting experiences, sharing knowledge, or helping people keep in touch with friends and family. In the process, they create communities—small and large—of those who write, read, and respond to them.[48]

My interviewees spoke of the emotional appeal of blogging:

> It's emotionally satisfying to put my words "on paper" or "on the web" in this case in an online journal. . . . I both read and write blogs, to keep in touch with friends and let them know what's going on in my life, if they choose to read about it. (Blogger2)

> The appeal to me is I have the chance to make my page the way I want to, I get to find out things of people that I never knew . . . when I go to the site I check my page and see if I need to add, change or do anything to it. (SocialNetworking9)

In short:

> You put up your ideas of what you think and feel. (Social-Networking9)

Blogs can be cathartic . . .

> I used to go on all the time to write in the blog, sort of like a diary. It's fun and gets your thoughts down and out of your head. (WorkGroup4)

> It makes me feel good, and it makes me want to say something important that I can't say out loud. If I have something I need to keep a secret I can say it there (on my page) . . . (AnimeLover)

and informative . . .

> I used to have a blog, just to mainly tell what I'm up to. . . . as a diary, just trying to track what I'm doing in the long-term view. (InfoGathering4)

As people express their feelings and share events and activities in blogs, they are in effect documenting their lives. Seemingly unconnected thoughts and experiences (and links to other people's thoughts and experiences) become connected, as we move from link to link to link, "hyperlinking" across the internet. A blog forms a story that can be no less than the story of a life; a story that can be updated, archived, accessed by others and linked to others—sometimes many others—at any time.

In the process of telling our stories and linking to and learning from others, multifaceted identities can take shape. Rebecca Blood describes discovering aspects of her own identity in the course of writing her blog:

> Shortly after I began producing Rebecca's Pocket, I noticed two side effects I had not expected. First, I discovered my own interests. I thought I knew what I was interested in, but after linking stories for a few months, I could see that I was much more interested in science, archaeology, and issues of injustice than I had realized. More importantly, I began to value more highly my own point of view. In composing my linktext every day I carefully considered my own opinions and ideas, and I began to feel that my perspective was unique and important.

Blood goes on to explain how such insights are gained through blogging:

> The blogger, by virtue of simply writing down whatever is on his mind, will be confronted with his own thoughts and opinions. Blogging every day, we will become a more confident writer. A community of 100 or 20 or 3 people may spring up around the public record of his thoughts. Being met with friendly voices, he may begin to experiment with longer forms of writing, to play with haiku, or to begin a creative project—one that he would have dismissed as being inconsequential or doubted he could complete only a few months ago . . . this awareness of his inner life may develop into a trust in his own perspective.

The readers of the blog—the community that springs up around it—contribute to the discussion and the process and may be inspired to begin their own "journey" of identity construction:

> [The blogger's] readers . . . will click back and forth between blogs and analyze each blogger's point of view in a multi-blog conversation, and form their own conclusions on the matter at hand. Reading the views of other ordinary people, they will

readily question and evaluate what is being said. Doing this, they may begin a similar journey of self-discovery and intellectual self-reliance.[49]

As blogs are linked to other blogs, a wide-ranging, nonlinear kind of conversation can begin. This occurs, in fact, in many kinds of online sites and documents, for in hyperlinking, people can trace ideas backward or forward and link to related ideas.

We use other online and mobile applications besides blogs to express ourselves. Discussion groups are often sites for the enactment of identity.[50] Text and instant messaging are popular forms of self-expression; some users say that members of certain of their social circles (like their families) would be surprised to see the texts they send and receive in others (say, with groups of friends), because such different sides of themselves are expressed in each group.[51] Still others report that owning a mobile phone (or not owning one) helps them define their identities, as does the fashioning and personalization of the phone and online icons.[52]

Portable technologies also permit us to act in a manner quite different from our offline demeanor if we wish, perhaps discovering directions in which to extend our identities. We can play different kinds of roles in each portable community, with, perhaps, no one the wiser. We can "play" more freely with our identities.[53] Just as Gamer1 notes that "my IM conversations often have a more bitingly funny tone than my standard conversations," we often act a bit differently in different settings, as different roles and tasks are required and enacted. Since we can visit so many places online and hide our physicality in them so easily, aspects of the self can now much more easily be experimented with, changed, and even manipulated.

In the portable community, physicality need not be destiny. Cultural studies scholar Sadie Plant points out that "access to the (computer) terminal is also access to resources which were once restricted to those with the right face, accent, race, sex, none of which need now be declared."[54] Without those social markers visible, we can alter our gender, race, sexuality, nationality, age. Physical traits can be disguised or re-created anew online. This allows us to explore what life might be like with another persona, or to explore different (or potential) aspects of our selves in ways that would be impossible offline.[55] Some women disguise their gender online in an attempt to avoid excessive or unwanted attention on that basis. But gender turns out to be difficult to completely disguise, especially over the long term. Women's and men's linguistic, semantic, and syntactic styles tend to be very different, and they have different interactional styles as well. Identity bending or switching, it should be added, is not without risk; people can and do get hurt when truths are exposed.[56]

Mostly, though, we don't switch genders or reshape our personae dramatically, not even in portable communities. More often, we use the luxury of anonymity and asynchrony to "fix up" parts of ourselves, to edit our interactions to make them just as we like. Some discover an enhanced inner strength:

> Maybe the thing I do best online is saying "No." I have a hard time looking at someone and saying no or talking to someone and saying no. (InfoGathering2)

or self-image . . .

> Participating in online groups I think honestly has made me a better person. (SocialNetworking8)

while others are more carefree . . .

> Telling people how you feel is always easier when you don't have to do it to their face because that fear of rejection or even fighting goes down. People become more carefree in what they say when using IMs or texting. (Blogger1)

and still others feel that they won't be judged as readily online . . .

> When IMing someone, I won't be judged for what I say. Or at least, I won't have to deal with their judgment because I'm not talking face-to-face with them. I feel I can be a little more open because I won't have to deal with their judgment of what it is I said. (TVFan1)

In myriad ways, people can "tweak" their identities a bit.

Some feel they can better disclose their "true" or inner self to others in portable communities than in face-to-face settings. They feel they are better able to present their "true selves" technologically and to have those selves accepted.[57] Gamer2 says that

> I am not a confrontational person but IMing helps me to be more real with what I have to say.

Some report that they feel better able to express their "real self" via texting than face-to-face or making voice phone calls.[58] And those who feel that their "true self" is best expressed over the internet are more likely than others to form close online relationships and, then, later, to move those friendships to a face-to-face basis.[59]

To be able to express one's true self in a portable community may never be more critical than when a person or group is threatened or marginalized in some way. If the internet offers "a unique opportunity for self-expression," as psychologist John Bargh and his coauthors point out, "then we would expect a person to use it first and foremost to express those aspects of self that he or she has the strongest need to express."[60] Many find online spaces to be safe (or at least somewhat safer) places in which to reveal and express identities that are not widely accepted in the societal mainstream. Racial and ethnic minorities, women, the gay community, the disabled, lonely, shy, or those with any kind of alternate lifestyle can all explore avenues for self-expression perhaps not available or as easily found in everyday life.[61]

Portable technologies can provide a means to "express and transmit often ostracized ideas and identities."[62] Supportive portable communities promote self-acceptance for those who feel marginalized and can help people feel more comfortable in all social settings, online and offline. Sociologist Douglas Schrock and his fellow researchers find that that an online support group for transgendered and cross-dressing individuals offers their members "a chance to excavate and validate a true self"—one that differs from "their born sex category."[63] Participants in this group describe their "joy at finding connection" in highly emotional terms. As one member put it:

> I just couldn't resist sharing my *elation* at having found you! . . . i stumbled across the group by accident during lunch today and my heart skipped, then skipped again. and again.

Another shares the strong sense of self found in the community.

> I told [the support group] my story. I was honest with myself and with the people in the room, and it was very cleansing. I felt really good. . . . I was amazed, it was like I broke through a shell. . . . It's almost like I had come home.

"Excavating" a self in this way is like being reborn into a new kind of family, Schrock and his colleagues conclude—one where people feel it is safe to be themselves.[64]

Marginalized both physically and socially, people with disabilities have much to gain from portable communities as well. They may feel they can present and express their selves more fully and spontaneously when "impairment is inaccessible to others' perceptual fields."[65] But there are limitations to this expression. The lack of disabled people in positions of power may serve the nondisabled status quo in ways that maintain and legitimize the oppression of disabled people. And people with visual or hearing impairments may be excluded from experiencing text, graphics, and audio; people with mobility impairments may not be able to manipulate the technology.[66] In general, those with disabilities tend to consider themselves a multidimensional, active group of people with many different personal experiences.[67] A study of disabled internet users in China found that for the minority of disabled people who do have access to the internet there, its use can lead to significantly improved frequency and quality of social interaction, and can reduce social barriers in both the physical and social environment.[68] For SupportGroup82, the benefit is in the support and social connectedness provided day and night:

> For someone like me who has health issues and limitations . . . my listserv is *always there* to answer any questions and you also get to know many through their posts . . . (emphasis added)

Physical limitations are generally less salient in sociomental than in physical space. Needed information can be accessed, social support found, critical interpersonal connections made, and different aspects of the self explored— all of which are particularly important to individuals in marginalized communities.[69]

Portable communities offer similar opportunities to explore aspects of the self to those who are shy . . .

> Talking to my friends online has enabled me to open up as a person. Sometimes it's easier to talk online than in person because I am sometimes shy. (MostlyEmail2)

> I find that I type things to my girlfriend that I would not normally say to her. I might be a little shy. (MobileUser8)

> Since I'm a very shy person, it's an easy way for me to interact with people. (MusicLover3)

and lonely . . .

> During the years right after my wife's death my listserv occupied most of my personal time as I had no immediate family, no nearby relatives, and was quite new in town. (SupportGroup3)

> At times that I am lonely or homesick I am able to easily chat with friends from home. (SocialNetworking2)

Of course, most of us feel shy and lonely sometimes. A text message or an email represents proof that we are not disconnected, and provides an easy, non-threatening means to connect in return.[70]

While it is dubious whether lonelier people are disproportionately drawn to use of the internet, it seems clear that those internet users that *are* lonely are generally more sociable online than they are offline.[71] Those who are lonely or socially anxious may feel better able to express their "real selves" in sociomental space. The mobile phone in particular has proven especially useful to those who are lonely or otherwise poorly connected, as it can confer "instant membership into a community."[72] Those who have a more difficult time establishing relationships face-to-face for one reason or another may have stronger need to use technology to express their selves and form connections with others.

Indeed, portable communities may ultimately be of the greatest use to people who experience challenges or barriers to conventional forms of socializing. The increased opportunities for social networking they offer may be especially helpful and gratifying for the shy, aged, disabled, or anyone who finds face-to-face interaction challenging. Portable communities can also help people who lack local community ties (such as couples without children) become more integrated into their neighborhoods,[73] and can help people from rural areas, or those who are housebound for some reason, find and venture into more accessible sociable spaces.[74] At the same time, it should be noted that spatially separated members of dangerous or destructive groups can now find one another more easily as well, and gain strength in numbers so as to espouse their views or harass and harm others (see chapter 9).[75]

Opportunities abound—for young and old, mainstream and marginalized—to create, exhibit, express, and extend their selves in portable communities. We seem drawn toward watching others display and express their selves, often becoming emotionally engaged with these others as we do. And in the process, our own selves continue to become ever more "fleshed out"—even when our actual flesh is hidden from view.

VOYEURISM, WATCHING, AND LURKING

Portable communities are windows into others' lives, lifestyles, and interests—windows into which many of us peek from time to time. Widely accessible and available social networking sites, blogs, discussion boards, and any of a number of sociomental spaces encourage both the creativity of their makers and the curiosity of their audiences. It is easy to become intrigued by the sheer amount of identity offerings displayed and to find some that catch our interest. We can find ourselves "identity probing," becoming involved in the lives of some of these distant others.[76] For there is an implied permission to "look" in the public nature of these displays. As Kevin Kelly writes about life online, "the act of making is the act of watching."[77]

Still, there is a voyeuristic element to peeking through windows. Whether we are probing online content or simply hanging out in online spaces, we often check out what others are doing, and thus can be seen, in a sense, as "voyeurs":

> I just look at what they have. I look at the clothes. (InformationGathering3)

> I visit to read other people's opinions on TV episodes and movies. (TVFan1)

> If I really have nothing to do and am really bored I might sign on and just look at pictures, pictures of famous people or my friends. (Lurker4)

Sometimes, it is hard to look away:

> At times, it becomes an obsession to check out what others are doing. (InstantMessaging2)

> It's fun to see what people are up to. I guess it makes me seem almost nosy, but I just like to see what people are doing. (SocialNetworking6)

And our desire to look at others from afar is also reflected in the popularity of internet- and mobile-accessed pornography (for more on this, see chapter 4).

Voyeurism sometimes takes the shape of lurking. Lurking—posting rarely if ever, but still regularly reading the postings on an online site (also discussed in chapter 4)—is an extremely common activity on the internet. Most of us lurk from time to time on websites, discussion boards, or blogs:

> I usually lurk—I find it interesting to learn from others. (SupportGroup1)

I usually lurk and see what people are discussing. (Music-Lover1)

I usually just lurk around other people's webpages reading their blogs and looking at their pictures . . . it's a way to stay in touch. (Lurker1)

Mostly I lurk . . . it keeps me learning new things, makes me more aware of all the people affected by this disease, helps me learn of other docs, stimulates me to be more proactive in my treatment and care. (SupportGroup7)

People lurk for any or several of the following reasons: because they want to be anonymous in their communities, preserve their privacy and safety, have work-related constraints, or find posting too time-consuming or burdensome. Though lurking may appear passive, it is actually a rather strategic activity which can satisfy many needs: to be part of an ongoing story or narrative, to be entertained and informed, and to find community and connectedness. All of this can be accomplished by simply being an interested part of one of the many audiences that comprise our portable communities.[78]

When we peek in on what others are doing online, we gain a sense of exactly who comprises our portable communities—what they are doing and what they think about and, most importantly, who they are, as individuals and as a collective. This "serves the essential purpose . . . of seeing and being seen," says sociologist and social network expert Duncan Watts.[79] We need to feel the presence of others in our networks and communities and to be present to them in return. Lurking helps us to feel one another's presence.

It is easy to peek in on home pages and blogs. While most people use blogs to document their lives for a comparatively few friends and family, others aim for the readership of a broader community. Either way, if these pages are publicly available, anyone can come across them and steal a glance. Blogs therefore facilitate the public expressions of the self and even offer the opportunity for a bit of exhibitionism: a platform, a "stage" on which to display the self to a theoretically limitless audience. Cameron Barrett says that blogs "are designed for an audience. They have a voice. They have a personality. In short, they are an interactive extension of who you are":[80]

I can choose my own design for my page and I feel like everything there is an extension of myself and my love of anime. . . . having these pages allows me to express myself to people who want to

> listen, and also listen to (read about) others who have the same
> passion as myself. (AnimeLover)

Others—potentially many others—may read one's words and view the content
of blogs and other kinds of sites. Through a feature by which people can
comment on others' postings, a "platform for dialogue between writer and
reader" is created.[81] As they respond to posts, members of audiences can
make themselves known, and conversations can result. These conversations
can inspire the formation of more intimate relationships and contribute
to the narrative of the community. "It was, and is, fascinating to see new
bloggers position themselves in this community, referencing and reacting
to those blogs they read most, their sidebar an affirmation of the tribe to
which they wish to belong," Rebecca Blood observes.[82] Slowly over time,
or perhaps rapidly, within days, a community—the "tribe" of which Blood
speaks—springs up.

Those who express themselves on these kinds of social sites are actually
communicating with a number of audiences both known and unknown.[83]
Sites may be geared to particular, specialized audiences, but users can not
know exactly who is "out there" watching, eavesdropping, participating.[84]
A number of unknown, unseen, people may be following, potentially even
caring about, the blogger or content creator. And even though "some users may sense their audience at a particular point in time," as boyd and
Heer observe, "they have no conception of who might have access to their
expressions later."[85]

Just as it can be a source of considerable unease or embarrassment to,
say, discover that a photograph or letter meant to be seen in one community
has been viewed by another, we must keep in mind that different audiences
exist online and that the content we post, though perhaps not specifically
intended for a particular group of people, may reach it. Young people are
sometimes horrified to realize that potential employers may view the photos
of them partying that they have posted on their social networking sites.
Most of us would not want our private letters (or emails) revealed to our
coworkers (or our parents). It is our task as content creators to negotiate
these numerous, diverse, unknown audiences, a task that requires complex
acts of projection and interpretation. Children and teenagers, among others,
may not be fully prepared to engage in such tasks.

"Watching" one another, therefore, and knowing that we are being
watched, is an active process of self-expression and other-interpretation. We
are detectives, of a sort, online, searching for clues as to who others (and
ourselves) really are.[86] We do this offline too, but must employ some particularly clever strategies to try to assess the veracity of online phenomena.
We look for specific kinds of identity clues, signals to one's identity that
are difficult to mimic. We may search for material online that has been

authored by an individual, or search a group's archives for certain old posts and compare them to newer ones, to see if the same person seems to have written both sets.[87] We may try to learn all we can from such indicators as email addresses,[88] names,[89] the links on a person's page,[90] linguistics and writing style,[91] buddy icons and IM away messages,[92] mobile phone ring tones and cases,[93] and even the timing of email messages and text messages.[94] We want to know our audiences, our online and mobile friends. Of course, as offline, this is not always easy to do.

Deception is always a concern in a community, and perhaps especially so in a portable community. We often call into question the authenticity of the personae we encounter:

> You can't expect the person on the other end of the internet to be what they say they are. (E-dieter)

> You're not always sure if the person you're talking to is really who they say they are . . . (MobileUser5)

The potential for deception, online and offline, is ever-present. However, as with one's gender, one's true self turns out to be difficult to disguise online, especially over time. Unless one successfully initiates a deliberate, sustained, unusually effective effort to deceive, one's real persona tends to seep through. While we may play around with aspects of our selves online, we are generally quite honest about our feelings, needs, and deeper impulses.[95] And it is generally the disclosure of feelings, not facts, that brings people together most intimately.[96] In addition, if we plan to meet an online friend face-to-face someday, we may fear the eventual exposure of lies and deceptions and therefore choose to be more truthful in our interactions from the start.[97]

Some consider the experience of "watching" others online less an act of voyeurism and more a genuine expression of emotional involvement. For Rebecca Blood, it is real emotion, not cheap voyeuristic glances, that is most likely to be exchanged when people blog. "There may be some people who follow blogs to 'watch,' but there are many others who really come to care about the lives of the bloggers," she states. "Many times, readers will come to a blog to read about a subject they are interested in, and slowly become invested in the everyday life of the writer as it is revealed in bits and pieces over weeks and months."[98] And even when we are merely lurking online we can be engaging with others in a deep and meaningful way. "Watching is doing," Goffman reminds us.[99]

In a multitude of social spaces, in truly unlimited ways, identities are created online. In surfing the web or merely following the links on a site, we can become easily interested in what we see—perhaps in a voyeuristic sense, perhaps becoming more emotionally invested. We know very little about voyeurism and potential voyeurism online; it is another area ripe for continued study. We do know that there is so much to see and do in these spaces that it has become common, especially among the young, to use portable technologies to multitask their activities and their social encounters. This can lead, I argue, to a kind of "hyperlinking" of the modern identity.

MULTITASKING AND THE "HYPERLINKING" OF IDENTITY

We each play many roles in a modern society—employee, partner, parent, son or daughter, citizen, friend, member of many social circles. Portable technologies in general, and mobile phones in particular, can facilitate the "harmonization" of these very different roles.[100] They can assist us in accomplishing a wide variety of tasks wherever we happen to be, minimizing unproductive time. It is common for busy, modern individuals to multitask, to do several different things at once, which is neatly paralleled by the way we also hyperlink online, clicking and moving among links and documents and pages.

In different portable communities, we do different things, talk with different people and audiences, and develop different sides of ourselves. Often, we do this in the somewhat disjointed fashion represented by hyperlinking. As we hyperlink online, moving from thing to thing to thing, we must learn to keep pace. Our brains, in a sense, must learn to hyperlink as well, juggling tasks and thoughts in a way that is more "weblike" than linear. There are no straight paths through the internet; it is not a novel to be read from beginning to end. The links we choose to follow take us on a special, individualized journey. No two people will follow a set of links, or take a hyperlinked journey through the internet, the same way.

Sending email, writing blogs, sharing videos, creating avatars, producing and consuming all kinds of online content—hyperlinking and expressing our selves along the way—the modern identity develops. Some claim that the modern technological individual consists of multiple selves, multiple personae, each of which is created in different settings, in response to different stimuli.[101] In this view, people are seen as consisting of multiple identities. Sherry Turkle proposes that we develop many aspects of our selves online that correspond to the many roles we play. In her view, we "cycle through" these aspects of self online, much as we click from one folder or file or message to another. Her analysis is worth quoting at length:

> We increasingly live in a world where you wake up as a lover, have breakfast as a mother, and drive to work as a lawyer. In the course of a day people go through dramatic transitions and it's apparent to them that they play multiple roles. Well-functioning people, successful people, happy people, have learned to work through all these roles, to cycle through them in productive and joyful ways. On the Internet you can see yourself functioning with seven windows open on your screen, literally assuming different personae in each of those seven windows, having all kinds of relationships, cycling among and being present to all of these roles simultaneously, having pieces of yourself left in these different windows as programs that you've written which represent you while you attend to another window. Your identity is a distributed presence across a series of windows. Increasingly, life on the screen offers a window onto how we are in our lives off the screen as well: we are people who cycle through aspects of self.[102]

This "cycling through" can be seen in the use of multiple usernames to explore and to "be" different personae, and in the adoption of multiple, different or alternate, sexual, racial, ethnic, age, or gender identities online.[103] But temporarily transgressing social boundaries such as gender or race does not dissolve them, nor does it make such categories meaningless, sociologist Lori Kendall reminds us; it can strengthen, even reify, them.[104] It is significant to note that most internet users speak more or less consistently with a single voice and username and exhibit a single identity overall.[105]

One could conclude summarily that the nature of the modern identity is multiple and that within each of us resides many distinct and separate selves. But there is also a certain singularity to identity—we have one body and are one person, and generally view our identity as an integrated whole.[106] While roles and identities are indeed more flexibly and fluidly enacted in portable communities (and elsewhere in the modern world), there is generally a limit to such flexibility.[107] Though we move among countless sociomental spaces and can experiment with and expand our identity in the process, we must still "anchor" ourselves in a single body. It is more helpful, I think, to think of the self as consisting of aspects or parts, each of which can be explored and developed as we interact with others, than to think of ourselves as consisting of distinct identities.

Georg Simmel writes of how we are each situated within a complex "web" of diverse group affiliations.[108] Taking on a large number of roles, and identifying, at least in part, with numerous groups, modern people can experience strain and conflict, but also a spectacular kind of freedom and flexibility of self-construction and expression. In the process, we have become more highly differentiated from one another, more different and specialized, than at any time in human history. In coordinating these diverse roles and expressing different aspects of ourselves in each group, the modern individual can become deeply complex and utterly unique. While this is an exciting, even freeing, proposal, it also presents a challenge: in creating a multifaceted self, we must find a way to be whole.

This is one of the central challenges of the modern age. Researchers in disciplines relevant to internet studies would be well served to examine how modern people negotiate these processes of self-construction and other-engagement and how they build and manage their overflowing portfolios of social connections. We must study the costs and benefits of doing so, the full range of consequences. Then, we must help people make the decisions that serve them best, with greater understanding of these processes and less fear and anxiety.

We do indeed "cycle though" roles and aspects of the self online. We are becoming used to hyperlinking and multitasking, both on and offline. Doing several or many things at once is cognitively (and sometimes physically) demanding and requires that we give only partial attention to each; it is a hallmark of our time. Moving smoothly and successfully among these aspects of the self to create a strong and solidified identity may seem a daunting challenge, especially to those of us who remember a pre-internet world and thus have known another way. Connecting with others technologically, and especially wirelessly and portably, is still a relatively recent phenomenon. These technologies have been integrated into our everyday lives in record time, but as might be expected, our understandings of them have not kept pace. Those of us who are older are caught between a landline and a portable world.

There is every reason to believe that upcoming generations of online and mobile connectors will find much of this neither daunting nor confusing. Modern individuals will always be challenged to create cohesive identities and strong communities and to understand the workings of both. But children and young adults growing up in a technology-rich environment have several advantages: skills in accessing information, interacting, and building social worlds online, in multitasking and moving unproblematically among social spheres, and in considering that which occurs in sociomental space to be very much "real."[109] They move more easily among online, mobile, and face-to-face contexts, carrying friendships from one domain to another smoothly because that is their world; it is all that they know. Technology consultant Rebecca Ryan describes an interview she had with a 17-year-old girl who had brought both a mobile phone and PDA with her to the interview:

> I asked her, "LaShonda, what do you think will be the impact of technology on the future of work?" She looked me in the eye and asked "What do you mean by technology?" I looked at all her gadgets on the table and asked, "Like this stuff!" She said, "*This is only technology for people who weren't raised with it.*"[110] (emphasis added)

The younger among us may not even experience a sharp distinction between online and offline phenomena, and it would not be unreasonable to speculate that in time the online-offline distinction may fade entirely. Angela Thomas claims that it is already happening: "For children, there is no such dichotomy of online and offline, or virtual and real—the digital is so much intertwined into their lives and psyche that the one is entirely enmeshed with the other."[111]

Anyway, the "virtual-real divide" is simply a "theoretical split," sociologists Brian Wilson and Michael Atkinson argue. It is a construct that does not adequately account for the way that we live our lives.[112] Modern individuals live their lives with much greater blending and blurring of experience than such a bifurcation suggests. As discussed in chapter 2, portable communities are *real* social worlds in which *real* experience takes place. Or as Nancy Baym says (in making what she calls her "favorite point"), there is no division between "real life" on the one hand and what she calls "virtual life" on the other. "It's always been true that what happens offline and what happens online interconnect," she says.[113] Indeed, the online-offline distinction may become less and less salient in time, and other systems of binary categorization (male/female, homosexual/heterosexual, black/white, work/leisure, producer/consumer) may become less and less useful too, as we create and live in complex technological societies.

In time, more and more of us may develop the "mental flexibility" and abstract mode of thought needed to appreciate complex identities and forms of relationship.[114] We may even collectively develop a kind of societal flexibility in thinking about social connectedness that I think would be optimal for understanding connectedness and result in a more nuanced understanding of social life. For the modern self is becoming more flexible and complex (albeit, generally, unified). There is great variety in the ways that individuals and groups enter into the process of constructing and expressing their selves. Portable communities support and encourage this variety, requiring and training us to develop highly differentiated, multifaceted selves. Future generations will be far better trained in this, and will, I hope, encourage the development of strong, multifaceted, free individuals who can tell the story of their society with increased clarity and understanding.

9

SHAPING A SOCIAL LANDSCAPE
Equalities, Inequalities, Possibilities

In this book, we have considered a number of issues regarding portable communities. We looked at how the community is built of cognitive connections, emotionality and intimacy, playfulness, and sociability. We examined the social networks that emerge as these connections are made and how people use them to make friends, find romance, work, learn, and get things done. Interpersonal interaction, we have seen, flourishes in portable communities, and is given color, texture, and "reality" by all of these internal social dynamics.

We then expanded our focus to look at some of the social dynamics surrounding the portable community in the society. We examined the constant availability of its members, the harnessing and control of social interaction, and the ways in which these communities influence the construction of modern identity. These societal shifts—along with others, and many others still to come—occur now at breakneck speed. The scope and pace of these changes can be startling, even scary. Perhaps understandably, many fear the outcome of these shifts.

In this chapter we widen our focus yet again, considering the larger (indeed largest) societal dynamics and addressing some of these fears and concerns. We look at how social divisions and power differentials, and several large-scale social problems—old and new—affect and are affected by portable technologies and communities. The role of individuals as they attempt to deal with these problems, sometimes discovering new reserves of energy and agency in the process, is also considered. Finally, the reciprocal relationship between technology and society is examined as we take a wide-angle view of a social landscape continually in flux, and consider the implications for society and humanity itself.

TECHNOLOGICAL DIVIDES AND POWER DIFFERENTIALS

Online and mobile technologies both create and replicate social power differentials. There is a pronounced difference and division between nations, regions, groups, and communities that have access to communication technology (the "information haves") and those that don't (the "information have-nots"). This is often called the *digital divide*.[1] Impoverished societies (located primarily in Africa, Central and South America, and Western Asia) lack not only wealth but sophisticated telecommunications infrastructures; they also have low rates of literacy. Political and bureaucratic impediments to the development of technology are rampant in these areas. Global technological use tends to increase with world-system status, democracy, cosmopolitanism, privatization, and competition in the telecommunications sector.[2]

Furthermore, much web content is in English, which can be exclusionary to those who speak other languages.[3] Embedded in cyberspace's technology is a bias toward the English language and other languages that use the Latin alphabet, teaches sociologist of cyberspace and cyberpolitics Tim Jordan. This has led to "the cultural domination of cyberspace by English languages that ensures some cultures feel excluded and marginalised and can make entry to many parts of cyberspace less attractive to non-English speakers."[4] Cultural norms online are often Anglo-American as well: heated discussions, competitiveness, and consumerism predominate in cyberspace (though electronically mediated discussions have actually been found to be relatively inclusive).[5]

The power to influence language and norms translates into the power to create a knowledge monopoly of sorts, in which, political scientist and cyberculture scholar Jodi Dean claims, "the interests of the marginalized and disenfranchised . . . are less likely to be served than those of the well financed and well connected, the wired and the savvy, who may benefit from a monopoly on certain ideas, images, and technologies."[6] Effectively closed out of many portable communities, and denied the information, resources, and influence that flow along their social networks, information have-nots (upward of four billion people worldwide) lack the means to substantially improve their situations. It becomes more difficult for impoverished nations to enter, let alone flourish within, a global marketplace.[7]

Mobile phones are more inclusive than are computers, although there are still regional and income-based differences in their adoption. Many of the same nations and cultures that are disenfranchised from internet use are not able to gain access to mobile phones. But lower start-up costs and ease of availability and use favor the diffusion of mobile telephony into some areas that have traditionally been telephonically isolated, as in parts of Bangladesh and Kenya.[8] There are not the exact same kinds of linguistic, literary, and skills barriers in mobile phone use as in internet use.

Even within such information-rich regions as North America, Europe, and Australia, the computer age is creating a new echelon of elites based upon technological sophistication, which, of course, correlates with money, status, education, and power. Entrance into portable communities is concentrated within wealthier, better-educated groups.[9] Latino and African American households in the United States lag considerably behind Caucasian households in computer ownership, and internet access is expanding most slowly at the lowest income levels (a contrast with European countries, in which diffusion of the internet is not so polarized by family income). In the United States, those who are more highly educated, economically privileged, English-speaking, and dwell in urban or suburban areas are more likely to engage in computer-mediated communication than their counterparts. Older individuals and the disabled are less likely to do so. The gender issue is more complex: internet users are increasingly female, though women and girls tend to be treated unequally as they build their computing skills, and they are more likely to question or doubt their skill levels.[10]

These differences, which echo other kinds of social inequalities, can be highly disenfranchising and preventative of economic and social advancement. Any minority use of the internet in the United States occurs within the overall social dominance of other groups.[11] Even as previously excluded populations obtain technological access, mainstream attitudes and stereotypes are apparent online and both can be exclusionary.[12] On the other hand, offline hierarchies may be being reshaped and augmented by online interaction because when people have the experience of building a community from the ground up they may have a greater belief in their own agency, their own ability to effect change.[13]

The sociological influences on technological use, then, are numerous and interconnected. In addition, there are physical and psychological influences on individuals' technological capacities and comfort levels. People need the means and motivation to use the technology.[14] In addition, some people use the internet or mobile phones haphazardly; their use is intermittent, or they stop altogether and start again several times.[15] The notion of a firmly bounded digital divide, then, may be too simplistic to capture the notion of technological exclusion.[16]

Clearly, though, those without access to computer and mobile technologies can not participate in portable communities nor fairly compete in a marketplace on which they are predicated (they can not take an online class, they can not access research or information via mobile technology, and their ability to gain assistance or contact a loved one in an emergency is diminished). Comfort with online and mobile technologies inspires more engagement with them and an opportunity to obtain high- (even medium-) tech jobs and to become fully integrated into technological organizations and environments. With society's increasing complexity, the minimum amount of information and skill necessary to participate in it is also rising.[17] It therefore becomes more and more difficult for people without the technology and the skills (and the ability to continually update one's technology and skills) to "catch up"—to become fully integrated into a technological society.[18] Though some say the digital divide is narrowing, it may be more accurately seen as *deepening*. That is, although more people may be gaining access to internet and especially mobile technologies, the chance for those who "have not" to become "haves" is decreasing. The deepening digital divide is becoming more and more difficult to cross.[19]

OLD PROBLEMS, NEW ANGLES

These are not new social divisions. They result in the exploitation of the powerless, hardly a novel occurrence. But it happens now in new ways. Social problems like theft, harassment, violence, and crime take on new dimensions when portable technologies are prevalent; their consequences potentially more dire when people are not physically copresent to one another, feel powerless, or when they feel like society is changing especially quickly.

Face-to-face accountability provides a check on certain types of human exploitation. It can be dangerous for people to be connected yet not be physically accountable to one another. Some crimes flourish in the face of anonymity, like stalking, pedophilia, the marketing of internet pornography to children, identity theft, drug trafficking, and certain forms of harassment.[20] These criminal activities proceed directly from the use of the internet and mobile phones to exploit others. It is difficult, for example, due to the decentralization of the internet, to halt the spread and purchase of pornographic images of children; this and the attendant abuse of children are now major issues in national and international law enforcement.[21] Controlling illegal drug trafficking on the internet and via mobile phones is also a law enforcement challenge, as rogue pharmacies, often overseas, are so difficult to locate and portable devices are often used for drug distribution.[22] Many legal experts consider current law to be insufficient to address these technology-based crimes.[23]

Governmental and organizational surveillance of individuals' activities and private communications is also a major concern and threat to individual freedom, personal privacy, and civil liberties. From work to commerce to travel, global surveillance has expanded significantly with the use of computerization to gather and store data of all kinds. Employers, governments, and organizations increasingly access and use these data for any number of purposes (for more on this, see chapter 6). Yet many of us feel relatively free online and do not focus on the extent to which our privacy and freedoms are diminished.

People can also now harass one another in new ways. They can take advantage of the anonymity and distance provided by portable technologies to do damage to people and reputations, frighten others, alter text and pictures in ways that might injure or embarrass others, or simply behave more rudely than they otherwise might. Sexual and other forms of harassment occurs in online contexts; widespread unwanted sexual attention, with substantial personal (and organizational) costs, has been reported.[24] Online and mobile technologies have been used to orchestrate crimes from gang rapes and drug deals to death threats and terrorism.[25]

Bullying via technology has become a major problem worldwide as well.[26] Harmful gossip, rumors, threats, and doctored photographs have been spread via internet and mobile device, and are of concern to many of my interviewees:

> Some people online just don't know when to give up . . . they just don't give up after you tell them you aren't up for meeting them and get really annoying and can become mean and scary at times. (MobileUser5)

> When it comes to writing things online, girls can be mean. They will say whatever they want. They don't care who it hurts or what they write and in the end it hurts a lot of people. It is the new easy way for girls to be mean to each other without having to say things face-to-face. (Blogger1)

Of course, danah boyd reminds us, bullying is a practice that can capitalizes on any available medium.[27] And this is true of many forms of online harassment. Still, as anonymity gives way to disinhibition and accountability, and legal recourse may be reduced, we see behaviors that might not have happened face-to-face take place in the dark corners of sociomental space.

Portable communities—like all communities—are also home to a host of behaviors that may not be illegal but are disruptive and bothersome. Flaming and trolling, as we have discussed, can be highly disruptive to the smooth operation of communities (although there are members who enjoy and even look forward such acts). These disruptions can be especially injurious to nonmainstream groups for whom group cohesion is especially important. And, of course, flaming can cause personal harm and distress in the form of harsh, hostile, pernicious (possibly illegal) remarks and threats. "People disrespect each other, and popular beliefs on the boards are supported almost to a racist extent," says WorkGroup5 of his experience in a work-centered community. Disruptions such as these can go so far as to shut a group down, or they may solidify it, as they force individuals to articulate explicit norms and rules and band together to banish the offending activity.[28] Or group norms may loosen and the members of a group or subculture may more freely indulge in previously discouraged behavior. Members of hate groups, for example, may be more likely to espouse their views online than offline, and can use portable communities to find one another more easily.[29]

The use of mobile phones in and of themselves can be bothersome. They are often used in formerly technology-restricted areas, like airplanes and restaurants, and at formerly technology-restricted times. We now interrupt face-to-face conversations to take mobile phone calls; we text message others in the midst of situations rich in physical copresence, like parties, dinners, sporting events, and classes. We have had to become accustomed to hearing conversations that would once have taken place in private occur nearly anyplace and anytime, the most annoying aspect of which, some say, is being forced to listen to only half of a conversation![30]

None of these problems are truly new. They have long existed in one form or another. But we experience them in new ways in portable communities. It is easy to forget about the public nature of online life while in the emotional, often intimate, moments of expressing ourselves online. We tend to treat cyberspace as a pretty safe place, and either forget the risks or assume they are minimal at the time. At the same time, social norms are changing. A certain degree of flaming, for example, is now both expected and accepted in online communities. And the intrusion of mobile phones into public spaces is increasingly tolerated, especially among the young.[31]

Still, there are dangers, as there always are when people form communities. As noted earlier, it is commonplace to "blame the technology" for things that happen in technological spaces. But technological determinism is an unfortunate response to these social problems, since it doesn't help us solve them. Technology has been used to help physically separated people get to know one another, discover commonalities, and share understandings since ancient tribes carved representations of their lives into the walls of caves for later passersby to find. Throughout time, photographs, letters, sound recordings, and the print and electronic media have brought spatially and temporally separated people together. But technologies often inspire worry, even panic, among their users, especially when newly introduced. It is not uncommon to blame technology for the social problems of the age.

A *moral panic* can ensue when such worry is taken to the extreme and a group attempts to exert moral control over another group or person in response. This occurs when a community identifies an external threat (sometimes one that reflects political or religious beliefs), there is a rapid build-up of worry and fear among its members, and an often extreme solution is proposed.[32] The popularity of social networking sites has inspired a degree of moral panic due to the presence on these sites of predators and bullies, both actual and potential. There is no doubt that the potential for harassment and crime exists on these sites, but in her in-depth study of the social networking site MySpace, danah boyd asserts that "there are more articles on predators on MySpace than there have been reported predators online."[33] She reminds us that moral panics are a common social reaction when teenagers engage in practices not understood by adult culture, and that moral panics have taken place in response to rock and roll music, jazz, television viewing, and even the reading of novels.[34] "The media, typically run by the parent generation, capitalizes on and spreads the fear with little regard for data or actual implications," boyd continues. "Examples are made out of delinquent youth, showing how the object of fear ruined them in some way or other. The message is clear—if you don't protect your kids from this evil, they too will suffer great harm to their minds, bodies or morals."[35]

Media attention is generally a component of moral panics. Whether "elite engineered" or a grassroots effort, a moral panic can take hold of nearly any community that feels sufficiently threatened.[36] To be sure, risks and dangers can be present in all kinds of sociomental spaces—they are very real. But in the throes of a moral panic, the scope of the problem may be misread, which is in itself a dangerous prospect. Usually, a panic recedes in time or results in social change. The offending activity becomes better understood or somehow integrated into everyday life. In the meanwhile, "tech-fear" can exert undesired effects. Jodi Dean describes how it disproportionately disadvantages less wired populations

> by deflecting attention away from web sites and Net uses that would enable them to make connections and find information and by keeping them technologically illiterate at a time when job, educational, and organizational opportunities increasingly require computer competence. But tech-fear has another, perhaps more dangerous, regulatory effect: it induces the need for a final authority amidst Net sprawl and information overload.[37]

In the absence of an "authority" that would subvert the decentralized nature of the internet, it remains the responsibility of all to remain alert to the dramatic, often exaggerated, quality of the moral panic. Online civil liberties and unregulated free expression must be balanced with the protection of minors and the safety of all. Models of governance that balance formal policing with more informal social controls are beginning to emerge within online social spaces.[38] Delicate balances like these can not be achieved without understanding the social dynamics inherent in the use of these technologies and the dynamics between technology and society.

People flock to social networking sites and portable communities in general because on some level they feel that the benefits outweigh the drawbacks. We must continue to study their social dynamics, their nature and effects, so we can assess and handle dangers fairly and appropriately and so we can guide our children (and ourselves) through these very real, very busy social spaces. "When you grow up the first thing your parents teach you is to look both ways before you cross the street and to not get in cars with strangers," MySpace cofounder Chris DeWolfe says, even as he points out the company's ongoing efforts to work with law enforcement to police and safeguard the site. "It's very similar for the internet."[39]

Portable technologies are merely the latest in a line of relationship-building and community-building media that facilitate interpersonal interaction from a distance. We use them because we need (indeed, crave) human connectedness and self-expression. But social connectedness will always have risks as well. We will hurt one another, mislead one another, and, inevitably, leave one another, whether by separation or death. As expert as we can become at harnessing social interaction, there is only so much of the process that can be controlled.

It is as dangerous to ignore social problems as it is unhelpful to overreact to them. Many of us remain unaware of potential technological problems and issues. Connecting technologically may *seem* to entail less social and psychological risk than face-to-face connecting.[40] But social interaction is always a risky business—and anonymous, invisible social interaction is perhaps all the more so.

There are both benefits and drawbacks to online and mobile connectedness and to all social interaction. Internet crime, harassment, social disruption, and surveillance are real and damaging; they are among the issues that a technological society must face squarely. Happily, the social networks that bring us together in friendship and work can also bring us together in addressing these issues. Increasingly, people are using online and mobile technologies to mobilize to bring about the kinds of societal changes they desire.

AGENCY AND ACTIVISM: MOBILIZING FOR SOCIAL CHANGE

Portable technologies offer interested participants many opportunities to mobilize their efforts for social change. Portable community networks are increasingly utilized for social activism—to rally people around a number of causes—jump-starting or providing needed energy to modern social movements. In using these networks, people can gain a sense of power and agency—of acting to make a difference in the world—a satisfying use of technology in a world in which social problems sometimes seem to overwhelm us.

Activism can take many forms, from direct action and protests to more long-term efforts to change laws and develop strategies for change. The success of these efforts depends on political opportunities, the strength of the group's collective identity, and the ability of the group to frame its cause and mobilize resources effectively.[41] Toward that end, the portable community can provide groups and individuals with an invaluable resource: a platform to reach numerous widespread audiences, a potentially loud, booming voice. In contrast to so-called "older" media like newspapers, radio, and pamphlets, multiple platforms on which social movements can be built (social networking sites, blogs, websites) are now easily accessed. "Don't hate the media," it has been said, "become the media."[42]

Blogs are now a significant political force; over a third of bloggers consider themselves journalists, challenging the notion that the traditional news media have a lock on deciding and producing what is news. While there is still a significant gap between the information haves and have-nots, many more people now have the opportunity to speak out in public forums. This is creating a kind of "civic journalism" that can potentially reach a very wide audience. With these types of outlets available to individuals—and *created by* individuals—people have considerably more power to speak and be heard than ever before.[43]

People are thus more likely to feel a sense of social power and agency and to become involved in efforts that could result in the changes they desire. This kind of power can be perceived and realized in positive or negative ways—or sometimes, a combination of both. In the following account, an online *hacker*—someone who enters private online spaces and accounts and seizes information illegally—describes how hacking represents to him an attempt to obtain something positive—social power and agency:

> This is our world now . . . the world of the electron and the
> switch, the beauty of the baud. We make use of a service

already existing without paying for what could be dirt-cheap if it wasn't run by profiteering gluttons, and you call us criminals. We explore . . . and you call us criminals. We seek after knowledge . . . and you call us criminals. We exist without skin colour, without nationality, without religious bias . . . and you call us criminals. You build atomic bombs, you wage wars, you murder, cheat and lie to us to make us believe that it's for our own good, yet we're the criminals. Yes, I'm a criminal. My crime is that of curiosity. My crime is that of judging people by what they say and think, not what they look like.[44]

Even those of us who do not illegally access online content can probably empathize with this yearning for free expression, as well as the impulse to judge and be judged by what we say, not what we look like. Portable communities are places where this ideal is often searched for and sometimes even achieved.

Online and mobile organization for social change is on the rise. The existence of trust and mutual understanding online and the ability to make decisions by consensus are all key to the success of organizing when people are not physically present to one another.[45] In addition, the existence of both "bridging" social ties (weak ties that link members of different social networks to one another) and "bonding" social ties (strong ties within the group) are critical to portable communities. People who belong to multiple portable communities effectively bridge those communities together, spreading information and agency from group to group. In fact, people who use the internet quite a bit and are part of several communities tend to attend more local activist events than others. The internet facilitates information gathering and political organizing among those who are engaged with the technology and the cause.[46] Those who are highly motivated to effect change, particularly those that oppose those who are in power and who believe that the government monitors their internet behavior, participate in politics online at the highest rates.[47]

Discovering and developing a strong resonance with other community members and with causes of interest are key to people's sense of themselves as active, involved agents in society. Cognitive agreement alone is not generally enough; agency must flow through emotions.[48] Emotional appeals have been critical components of such social movements as the civil rights movement (in which speeches were targeted specifically to the emotions of recruits and supporters), the women's movement (one of the aims of which was to prepare women for the emotional fallout of resisting dominant paradigms), and the gay movement (in which the fostering of pride was a critical element). The exchange of emotion—which, as we have seen, is in abundant supply in portable communities—seems to be a precondition for influential consciousness-raising.[49]

The internet, and blogs and social networking sites in particular, can be thought of as a kind of commons: a social sphere well suited to consciousness-raising and political expression. Political and special interest sites and online civic messaging are prevalent. Politically oriented blogs—many of them extremely influential—proliferate; some even challenging the traditional news media as credible information sources. A wide variety of political or opinion blogs famously create "buzz"; they take local stories and make them national, generate discussion, and drum up support for people and issues. They can set a more widespread social or political agenda, for if certain blogs are discussing an issue, the mainstream media will often start addressing it as well.[50] If nothing else, political blogs have inspired a heightened "rights consciousness" among those who read them:[51]

> Online groups have allowed me to not only be more open to others and what they think but also to be more informed about worldly matters. (InfoGathering5)

> My participation in "free speech" communities has been beneficial. It has shown that there are people out there with different outlooks on life than mine. (InfoGathering6)

Whether or not the stated purpose of the portable community is political, once people find one another and begin to get to know one another, they may find themselves inspiring one another to act in new and unexpected ways.

People are often surprised by the extent to which they find themselves motivated to work for change within portable communities, as was Gamer1:

> I have become unexpectedly galvanized on a lot of issues that are off the radar for most people. I have joined the Electronic Frontier Foundation, and I denounced a congressman on the radio for sponsoring legislation that will restrict innovation.

And consider the story of this young man, a junior in a California high school who helped lead a student walkout in early 2006 that saw thousands of students leave classes to protest proposed anti-immigrant legislation. I excerpt his very interesting account at some length:

Well, it all began on Sunday. Sunday morning, like around 9:00. I was actually on MySpace. That's the website that was a real big help in getting the word out. And I seen a bulletin that a girl from L.A. had posted up; and it showed, you know, the march—the rally that L.A. had and the hundreds of thousands of people that showed up. And it also had some sort of a slideshow. And it showed these pictures of, you know, the rally and migrant workers and everything. And I guess that that was the main thing that kind of got my attention and said, well, if California, you know, is trying so hard and people around the U.S. is trying so hard—because you hear about the news, you know, people marching in other states—and I said, well, we should try to do something here. And she had actually posted up a national walkout for Monday the 27th.

And I guess usually when you see those type of things, you know, walkouts, nobody really does them; but on those, I guess, you know, it's for a reason, it's for a good cause. I decided to help out in any way that I could; and, I mean, immediately I had to call a best friend of mine, Miguel, and as soon as we called him, I mean, we spent the rest of the day just calling, texting people on their cell phones, emails, and by anywhere. We got some—I had made some flyers, and I made some that same night. And, I mean, it was just a one-day thing that we hoped to try to get as many people as we could. . . .

A lot of people were like, "Well, it's too early, you know." People were like, "Okay, okay. We'll do it. We'll do it. We'll walk out. But when?" And I was like, "Tomorrow." And they're like, "Ahh, you know, I don't know if we're going to get that many people. I don't know if we're going to be able to get the word out." And it was just one of those things where I was like, "Well, just that much—" you know, just—I was empha-sizing to them, 'Well, as little time as we've got, that should, you know—You should be able to help me out more and just call anybody you know. I mean, call your whole phone book in your cell phone. Email everybody, all your contacts." And, I mean, it was—in the end, everybody was for it . . .

The main focus that we had on everything that we sent out was: as soon as you get this information and know about this, tell someone else. So, it was one of those things, like a spider web that just kept growing, a network of people that just kept telling more and more people. And I guess that's what really helped out spreading the word in all of this.[52]

In the course of one calendar day, utilizing the internet, cell phones, and a big stack of old-fashioned flyers, this young man was able to explicitly utilize his social networks—and the people in them who had "bridging ties" to other networks—to mobilize an enormously successful walkout. People who might otherwise have failed to become involved could become highly, and almost instantly, motivated to action.

Efficient, user-friendly, and critical to disseminating certain types of information to those that need it, portable technologies have become indispensable tools for large-scale political organization. The 1999 Carnival Against Capital! demonstration that inspired hundreds of thousands to protest neoliberal institutions and their related globalization policies primarily (and covertly) used the internet to organize.[53] An April 2006 series of marches and rallies in support of human rights drew over one million people across the United States and was organized in large part via social networking sites.[54] On an even larger scale, the group of liberal bloggers and online activists often called the "netroots" raised millions of dollars for Howard Dean's 2004 presidential campaign and many congressional and senatorial candidates since, and may total over six million people.[55] Barack Obama's 2008 presidential campaign was built and funded on a groundswell of over a million internet supporters. And an advocacy blog has been credited with helping to incite and sustain a major revolution in the former Soviet republic of Kyrgyzstan. It had become a unique, critical source of information not available from other sources.[56]

Today, Kahn and Kellner find that

> broad-based, populist political spectacles have become the norm, thanks to an evolving sense of the way in which the Internet may be deployed in a democratic and emancipatory manner by a growing planetary citizenry that is using the new media to become informed, to inform others, and to construct new social and political relations. . . . The global Internet, then, is creating the base and the basis for an unparalleled worldwide anti-war/pro-peace and social justice movement during a time of terrorism, war, and intense political struggle.[57]

New social movements, organized and coordinated through both portable and traditional technologies, are bringing together coalitions including labor, environmentalist, feminist, peace, and global social justice activists. "Social movement theory has typically focused on local structures, leadership, recruitment, political opportunities, and strategies from framing issues to orchestrating protests," says sociologist Lauren Langman. "While this tradition still offers valuable insights, we need to examine unique aspects of globalization that prompt such mobilizations, as well as their democratic methods of participatory organization and clever use of electronic media."[58] As our society becomes more global in structure and impact, we must consider how technology generates social and political participation that could potentially be global in its reach.[59]

It is possible, though not ultimately most effective, to be an "online activist" only; to work for social change solely within portable communities. Some never meet with members of their activist communities face-to-face:

> I am part of two email lists that go out daily on political and community events, to which I devote 30 minutes to 1 hour per day. . . . I update myself on current events and network and communicate with political people, without leaving the house. (Activist)

> I'm a feminist activist. We're a long way from equality and since I'm not well enough to travel and organize much, I try to work online for change or just to stop the Republican onslaught. . . . we communicate almost exclusively online. (SupportGroup4)

> I have a feeling of power in that I can contact legislators and work with groups to influence legislation, all online. (SupportGroup8)

Technology can indeed be empowering, though it is not without its constraints. But technological outreach alone is not enough to create real and lasting social change.

"If the internet can assist people seeking progressive social change," suggest cyberactivist scholars Martha McCaughey and Michael Ayers, "it will do so not as an inevitability, not as a cause but as a means of change alongside other forces."[60] Or, as journalist Perry Bacon argues, "You can't move elections with just modems and IM."[61] To maximize the effectiveness of a social movement, at least some of its members must gather face-to-face. They may publicize rallies and marches via online and mobile technologies, and use portable social networks to recruit and raise money, but to make the greatest difference they must canvass neighborhoods, go to the polls, take to the streets. As David Silver says regarding a website that had organized a series of rallies on immigration, "What I like about [the site] is its ability to help users go from virtual to physical. One click on the map gives users the ability to find a march or marches near them. Thus, they a) visit a web site in order to b) participate in a march."[62] As we have seen in other contexts, the online-offline leap is a critical one to make. Ideally, online and offline communities will interpenetrate and complement one another as people seek to exercise their agency in ways that empower them to create lasting social change.

A LOOK AHEAD

Throughout this book I have looked at the internal and external social dynamics that surround portable communities and the technologies used to build them. As always, though, when studying something perpetually in flux—whether that be social interaction in general or the ways that people find to form relationships, communities, and a social culture—one's findings can never be more than tentative. There are inevitably more questions raised than answers provided. As a researcher and as a learner, that is both exciting and frustrating. There is much to be done, which is exciting—there are many excellent researchers and scholars in fields related to communication and information technologies undergoing relevant work—but the fact that it will never be anywhere near conclusive can be frustrating as well. To study ever-changing technologies and societies is like trying to take a tape measure to a river: the best one can do is to examine it as comprehensively as possible at a given point in time, knowing full well that tomorrow the currents will have shifted a little or a lot.

This book has charted some of these social currents. Not only are there others to explore, but those examined here will ebb and flow with time. One thing is certain: those looking for research projects in the fields of sociology, psychology, communication, information science, media studies, computer and internet and technology studies and all related disciplines, now too many to list, will find them here. Solid, creative research and theory regarding the relationship between technology and society has never been more needed. Interdisciplinary (indeed, transdisciplinary) research exploring the intersections between and among these varied areas is much needed, especially as so many of the topics most critical to explore are not the province of any one discipline. These include further (and ongoing) examinations of technological connectivity and interactivity; issues of design, access, and status and power differentials; and the rhetoric, language, and discourse of cyberculture.[63] Throughout this book, I have identified other, more specific topics that are most acutely, in my opinion, in need of further study. But all inquiry that will help us understand the social impact of communication technology, critically and comprehensively, is up for grabs. There is plenty of work to do, and there always will be, for you can never step in the same river—or internet—twice.

Given the impossibility of finishing the job, then, what are we to make of all of this now? What can we say—at this moment, anyway—about a society in which portable technologies and communities are so prevalent? What of a future in which their influence is certain to be increasingly prominent?

We now use technology in highly creative and participatory ways to build social culture. We use portable technologies to interact with one another almost everywhere we go and almost every chance we get. We do this as we travel in social networks that we have collectively created and contributed to and we do it even when we have never met the others in the network, and will never meet them, face-to-face. Most of us are a part of numerous large, active networks that consist of both sociomental contacts and people we see face-to-face as well. As these networks criss-cross and become knitted together, they provide a kind of scaffolding upon which a society is built; a society that extends indefinitely and infinitely across space and throughout time.

Portable technologies are used to establish these social networks and link them together. They have changed the ways that people think, feel, play, hang out, shop, work, fall in love, and organize for social change. They have spurred an impulse for creative, social, and, in some cases, political participation. In doing so, they enable—link by hyperlink, network by network—the construction of society. The social uses of portable technologies, it seems certain, will only continue to increase. We have taken "only the first steps in this great shift from audience to participants," Kevin Kelly notes, "but that is where it will go . . ."[64]

Exactly what we decide to create, talk about, and share, of course, is up to us—individually and collectively. But the shift of which Kelly speaks is underway and has the potential to be seismic. More of us have a chance to speak and to be heard by unlimited others than at any time in human history. "We're looking at an explosion of productivity and innovation, and it's just getting started, as millions of minds that would otherwise have drowned in obscurity get backhauled into the global intellectual economy," says Lev Grossman. This has created an opportunity for us as a society to build no less than "a new kind of international understanding, not politician to politician, great man to great man, but citizen to citizen, person to person."[65]

Still, there are dangers, risks, and constraining forces. Too many of us are excluded from technological access and the opportunities to fully participate in a technological society. Too many are harmed and harassed. Surveillance, crime, and reduced civil liberties may result in a restructuring or collapse of parts of the system or a gigantic weakening in the rights of the individual. To use online and mobile technologies to build a social culture is really nothing less than a grand social experiment, the outcome of which, like all good experiments, is uncertain.[66]

Future societal prospects of technology use, John Bargh reminds us, are neither necessarily grave nor liberating.[67] Portable technologies can (and will) be used in any number of ways for any number of purposes, bringing us to any number of sociomental destinations. The characteristics, motivations, goals, and behaviors of those who use them will characterize the journeys we take with these technologies, now and always. Though it is irresistible to make predictions and speculations regarding the road ahead, it is also impossible to expect such predictions to hold firm, especially in a time of such relentless change.

But surely social interactivity, creativity, consumption, and hands-on participation in communities large and small will remain a prime use of portable technologies. We reveal and express ourselves and create communities and relationships in response to the deepest of human needs. Simply, incessantly, we reach out to one another. It should come as no surprise that when technology helps us do this with relative ease and frequency, it should become indispensable, even cherished (try separating a teenager from his or her cell phone!).

As we have seen, younger people who have grown up immersed in technology, accustomed to carrying it with them and on them all the time, may be far ahead of the rest of us both in terms of skills and conceptual understandings. Using technology so frequently and fluently that they may not even consider it "technology," the lines between public and private, work and leisure, the real and the unreal, and a number of other social categories, may be becoming utterly and permanently blurred. "Their worlds bleed together," observes information technology professional Charles Grantham of young technology users. "It is pretty useless to draw borders around different spheres of life for them."[68]

In multitasking and hyperlinking, we are becoming used to doing things in a somewhat more fluid, less linear, fashion. Our identities follow suit: they are more highly specialized and multifaceted than at any point in the past, a trend we can expect to continue. At the same time, there are losses incurred that surely affect us in ways we can not even yet see: losses of privacy, of face-to-face accountability, of time for reflection and *disconnection*. But not, I argue strenuously, a loss of community. Relationships and communities are formed in different ways online than offline, but they are out there in abundance and they are not necessarily *less than*.

Some worry that social solidarity and the general stability of our society are endangered in a society driven by continuous technological innovation and adoption. Civic society and community in general is often pronounced dead or dying.[69] As it turns out, online and mobile technologies provide opportunities "for the disaffiliated to reconnect" and "for the super-connected to shy away"[70]—and everything in between. Portable communities provide infinite means for people to address their needs, desires, and potentials. We use these technologies to make dates to get together face-to-face just as often (if not more often) than when we use them to connect with those whom we never see. And we use them in combination with one another to remain in contact with an increasing number of ever-widening social networks.

Portable communities do not seem to replace but rather to supplement and even make possible face-to-face encounters. Though danger will always be part of connecting technologically—for risks and dangers are found in all forms of interpersonal interaction—it is also instructive to note that the more technologies individuals use, the more help and assistance they tend to receive when trouble strikes.[71] In short, portable communities have the potential to enhance social relations of all types through prompting social contact and the creation and use of strong, useful, active social networks that meet both in sociomental space and in the physical world.

Certainly, some of us become excessively immersed in online and mobile connecting. And it is not uncommon to forego some face-to-face contact in the course of using them (though it should be remembered that face-to-face activities are not in and of themselves always, inherently, satisfying).[72] Anyone who spends significant time engaged with others via a technological device must shift at least some time and effort away from other activities.[73] Many of the activities displaced by the internet, however, are marginal to face-to-face interaction: they include watching television, talking on the telephone, and sleeping.[74] Though some users may become more isolated or socially alienated in the use of technology, the opposite is more often true. Online and mobile connectors generally become more deeply and fully integrated into society, both offline and online.[75]

Life in the portable community is not the same as or equivalent to or substitutable for face-to-face interaction. In sociomental space, the senses are not as fully engaged, people can not care for one another in as wide a range of ways, and many of the satisfactions to be gained in face-to-face relating can not be achieved. The idea of people spending weeks, months, years on end connecting only via computer or mobile technology—living, loving, *parenting* solely online?—seems bleak and sad, not the world most of us want to live in. But it is *not* a realistic assessment of where we are headed at this time. The face-to-face bond—the human touch—remains the bedrock of society.

But physical contact is simply not all that enriches us. Portable communities, as the people I interviewed told me time and time again, have deep meaning for people, facilitating social connections that can be strong, authentic, and accessed wherever and whenever they like. There are also very real consequences (the classic test of the "realness" of social phenomena).[76] When portable communities are thought of or treated as or implied to be fake or suspect, those of us who participate in them are inevitably, and needlessly, diminished.

My interview subjects all told me that they form real, consequential social connections and genuine communities in the use of online and mobile technologies. They do this easily, routinely, as a matter of course. Barry Wellman and Caroline Haythornthwaite tell us that we should no longer view the internet as something apart from our lives or as "a bright light shining above everyday concerns," but as a taken-for-granted, integral part of our existence.[77] Someday—possibly tomorrow, possibly today—the youngest among us may not even make a distinction between life online and offline. For children, as for many adults as well, time spent in portable communities may be becoming remarkably . . . unremarkable.

Those with technological access and sufficient skill and interest can become highly involved, even immersed, in portable communities. The dynamics of sociomental connectedness—cognitive resonance, emotionality, supportiveness, intimacy, fun, sociability, practicality, convenience—inexorably draw us in. The constant availability of others, our ability to harness and control social interaction, and the opportunities afforded us to express ourselves and create identities, intrigue us—even as they may cause us anxiety or harm. In a technological age, "things are becoming more complicated and lively," Wellman and his research team conclude.[78] We do not know exactly how social life will change, but that it will remain both complicated and lively is a safe prediction.

For social connectedness is now firmly, irreversibly, portable. Our portable networks and communities provide us with friends and lovers, power and resources, that we take with us wherever we go. At the same time, they perpetuate social inequality, crime, and other social problems. Members of technological societies have always used those technologies available to them to bridge physical distances and to get to know faraway others. Whether the associations that result proceed positively or negatively—for good or evil, for fun or profit—social interaction has always depended at least in part on technological mediation.

Technology and society are enmeshed in a kind of incestuous, reciprocal relationship with one another. New technologies emerge and are appropriated by us—often in ways that could not have been predicted—to meet our needs. Our need for sociation is so great that we frequently, portably use these technologies to help us get to know and interact with others. Social networks, communities, cultures, and whole societies are built on the scaffolding of these technologically mediated interactions. And in a kind of giant feedback loop, new technologies and applications that then help people meet their interactional needs and desires (or somehow exploit such desires) are invented. Previously unimaginable forms of interaction (landing a job online, achieving sexual pleasure by phone, exchanging worries and successes with people a continent away) become more desirable, more imaginable.

In this way, technology and society shape one another. Technologies help us form (and dissolve) relationships and networks, create spaces in which to hang out and play, manage and organize an unprecedented number of social connections, and create selves and societies more richly textured and complex than ever before. We do not and can not know how these potentialities will be realized, or what form our future societies, communities, and identities will take. But we can expect that the social dynamics in and around these portable communities will remain complicated and lively. For above all else, society is endlessly and insistently dynamic—pulsating with change and alive with possibility.

ACKNOWLEDGMENTS

Writing is one of my favorite activities: it is rarely hard for me to find or make the time. But the many related activities involved in undertaking a project like this—designing and executing the research, collating and reviewing the literature, editing, proofreading, and indexing—are another matter. Those tasks are easier for me to sidestep when teaching, sports, friends, family, music, and fun (not necessarily in that order!) beckon. I must credit, then, sociologist Donna Huse for encouraging me so passionately to write this book. In a casual conversation with Donna a few years ago, I found myself listing the litany of reasons why taking on this project just then would be all but impossible. Donna and I barely knew one another, although she had reviewed my first book and knew my work well, but as we talked I noticed her becoming just this side of livid at my laziness. "Write it," she said with an urgency I had not felt in myself in some time. "Just write it. Now."

For some reason, Donna's words took hold. I heard them in my head constantly during the many months and many tasks involved in writing this book. Since that conversation, I have been fortunate to receive the encouragement of many, but it is Donna's voice I heard the loudest during the longest of the hours spent bringing this project to life. I thank her here, then, with my deepest gratitude, for her words and for somehow getting me to hear them.

The College of Saint Elizabeth, my professional home, has warmly and actively supported me as a scholar. A more enjoyable and collegial place to work I can not imagine. Thanks especially to Sister Francis Raftery for the leadership, the laughs, and sending me home with food after every meeting, and to Dr. James Dlugos, Dr. Carol Strobeck, and Sister Ellen Desmond for encouragement that was genuine and gentle. To my dear students, you know how much I adore you, and to Laura Napolitano,

Sharifa Extavour, and Samia Canzonieri, I thank you in a special way for your painstaking and invaluable research assistance.

The 87 anonymous individuals who shared with me the stories of their social connections and communities are the backbone of this book. Without their generosity of time and thought I would not have known what I had to say or where this book was headed. To all of you—and you know who you are—my heartfelt thanks.

Many professional colleagues in a variety of settings and disciplines contributed to this project. They include the following, whose personal exchanges with me resulted in critical insights that go far beyond their published works and have enhanced this work greatly: Barry Wellman, Christena Nippert-Eng, Karen Cerulo, Sherry Turkle, Jim Katz, danah boyd, David Silver, Rich Ling, Eszter Hargittai, Nancy Baym, Lori Kendall, Rebecca Blood, Rick Eckstein, Monica Nicosia, Tracy Budd, and Jeanie Akamanti. And I will always be grateful to the longtime support of my mentors Eviatar Zerubavel, Ira Cohen, and Karen Cerulo, and my colleagues and friends at the College of Saint Elizabeth, Rutgers University, and elsewhere—too many to possibly name.

To the fine editorial, production, and marketing team at the State University of New York Press, especially Nancy Ellegate, Eileen Meehan, Fran Keneston, and Dana Foote, many thanks for a job expertly done. I thank the brilliant poet Joseph Millar for the words that grace this book's opening epigraph. They come from his stunning tribute to the beauty, pathos, and dignity of everyday work and workers, *Overtime* (2001, Eastern Washington University Press). I also thank the gifted artist Raul Villareal for the cover concept and execution.

My deepest thanks and love go to my family. My brother John Chayko and his wife Claudia Scotti, my sister Catherine Chayko and her husband Gary Wassmer, and my devoted mother Terri Chayko comprise surely the best (and wittiest) portable community of all time. Our regular email "circle" has proven a great test case for my ideas and theories, and has made the miles between us fall away. And most of all I thank my husband and partner Glenn Crooks, who puts up with my moods, quirks, and solo weeklong trips to the beach ("to write, of course!"), and my children Morgan and Ryan, who inspire and cheer me on constantly. Almost every day while I was working on this book, Morgan would ask me, "Mommy, how many pages did you make?" Finally, I can tell her, "All of them!"

Appendix I

THE METHODOLOGY

People who spend a good deal of time online are generally comfortable there.[1] For this study, that is where I met them; I interviewed them in the very sociomental spaces they so often inhabit. I also reached out to professional colleagues who conduct research on internet and mobile connectedness in a number of disciplines, asking them to elaborate on various aspects of their work, seeking further insights into certain topics and concepts, and incorporating these insights, along with those gained in the process of researching my prior book, into my understanding of portable social connectedness. And I performed an extensive review of literature in a number of fields relevant to internet and technological studies, particularly those in which I am trained: communication, psychology, and the one in which I teach and am most active, sociology. Using these methodological strategies, I present the claims and ideas put forth here not as the definitive or "final word" on any aspect of internet or mobile connectedness but as a reasoned and scholarly interpretation of a phenomenon, limited by scope and by the passing of time—a look at portable communities at what is, relatively speaking, a moment in time.

The methodology for the empirical portion of this study could best be described as an "electronic interview."[2] I conducted 87 qualitative online interviews. In them, I described the project and the kinds of social dynamics I was exploring and then asked a series of related questions. Though conducted online, these interviews are very different from quantitative online surveys or questionnaires: these are in-depth, semi-structured, open-ended, multi-phase interviews, permitting give-and-take between researcher and subject.[3] My intention was to gain insight into people's states of mind and feelings regarding online and mobile connectedness and their participation in portable communities.

There are several modes of conducting online research, each with strengths and weaknesses: synchronous modes tend to produce a chatty,

205

spontaneous quality, while asynchronous modes (like the current study) tend to result in more considered and thoughtful responses. I chose email as the platform for delivering the questions and conducting the research, as it lends itself to private, considered conversation and a sense of one-to-one interaction.[4] Electronic interviewing has been determined to be a viable and very effective means of gathering information online, especially when the topic at hand is online life and behavior, for in a very real sense the interviewer is meeting the subject on his or her own "turf."[5]

Just as relationships online can be just as intimate as those transacted face-to-face (if not more so), so can electronic interviewing, if carefully designed and delivered, foster a close and indeed intimate interactive experience. Asking and answering questions by email allows interviewees plenty of time to consider the issues involved and give thoughtful responses, increasing the potential depth of understanding for both parties. Neither the researchers nor the interviewees need schedule appointments or be concerned with the effects of interruptions. It is also less likely that subjects will give responses in accordance with the expectations of a perceived evaluator, as telltale nonverbal gestures and signals unwittingly given off by the interviewer are absent online.[6] Other advantages of email interviewing include

> reduced time and cost; convenience; unimportance of geographic location and the possibility for more sampling diversity; the potential for large amounts of data to be accumulated quickly; the allowance for more thorough and thoughtful follow-up and clarification; single-step, non-interfering recording and transcription; no danger or discomfort for the researcher; and the ability to continue the interview process until the researcher is satisfied that a saturation point has been reached.[7]

"Many of the difficulties inherent in face-to-face interviews are overcome in the electronic medium," educator and online researcher Kay Persichitte points out.[8]

On the minus side, it is not possible to verify the identity of respondents online, which can impact the potential validity of responses. The lack of visual cues can hinder the transmission of certain understandings. Email interviewing is less spontaneous and does not flow in the same way as a face-to-face interview, though that can be compensated for by the additional time for reflection and depth that asynchronous communication permits.[9] Finally, some people might prefer to share some types of information face-to-face rather than via email (though others may prefer the anonymity of email). It is likely, however, that people who are involved members of portable communities are fairly comfortable sharing their stories and experiences over the internet, and the promise of confidentiality and anonymity by a researcher at an accredited institution of higher learning seems to have satisfied subjects concerned about such issues and therefore enhances the validity of the results.

To design and conduct qualitative research in a valid and credible way is always a challenge, as calculable levels of significance such as those determined using quantitative methodologies are not appropriate and do not exist. It is then all the more important that qualitative researchers adhere to standards for quality practice so that their arguments convince and findings are valued. Noted internet researcher Nancy Baym has identified six criteria for "quality" qualitative internet research that concerns the internet. They are: the research should be grounded in theory and data, demonstrate rigor in data collection and analysis, use multiple strategies in data gathering, take into account the perspective of the participants, demonstrate awareness and self-reflexivity regarding the research process, and take into consideration interconnections between the internet and the lifeworld in which it is situated.[10] The present study fulfills each of these criteria, especially in its grounding in theory and data and accumulation of insights, information, and literature from numerous experts in many related fields. Furthermore, information has been provided regarding all interview subjects so that their perspectives can be placed in context (profiles of each interview subject are provided in appendix 2).

To select the subjects whom I would interview, I used a "snowball" sampling technique very similar to that I employed for the research for my first book, *Connecting*. This form of strategic informant sampling is generally used in studying specific aspects of people's experience and of the groups and societies of which the subjects are a part. The strength of this particular type of sampling is the identification and characterization of social forms and types of interaction. In addition, each subject can provide a depth and breadth of detail regarding topics and questions of interest impossible in large-scale quantitative surveying. Its weakness, the weakness of all qualitative methodology, is that results can not be universally generalizable but are, rather, illustrative of concepts, ideas, and constructs.

In strategic informant sampling, the interviewer builds up a sample by hand selecting a small number of people to interview and then asking those initial subjects to supply names of other potential subjects who would be appropriate for the study. From there, the pool of interview subjects "snowballs." For this study, I began in April 2005 with a set of six subjects whom colleagues at several U.S. colleges and universities referred to me as appropriate since they were a member of one or more portable communities and might be inclined to discuss the experience with me. I emailed the consent form (including the project's IRB approval from the College of Saint Elizabeth) and a description of the study to these individuals. If they were interested in participating and indicated their consent, I sent them an initial set of six questions. In return emails, I thanked them for their participation and asked from zero to ten additional questions (depending on whether and how the issues they raised may have called for additional probing). As the project progressed throughout 2006 and 2007, the number of initial questions I asked subjects increased to ten, as new issues emerged during the course of the research and became important to the study, requiring additional data and clarifications of responses. I then revisited my interviews with my earlier subjects and asked them the additional questions that had arisen. Together, my subjects and I created what has been called a researcher-subject "feedback loop."[11] When technological and relational skills are strong, electronic interviewing can be particularly effective.[12]

I asked my initial set of six individuals to refer me to other individuals who met the criteria of the study. Additionally, when I encountered groups of people whom I knew were involved in portable communities (after a public talk or lecture on the topic of online connectedness, for example), I recruited potential interview subjects. I then followed up with those who demonstrated continued interest. In all, 87 respondents completed the electronic interview.

The interview consisted of wide-ranging, multipart questions about which respondents were encouraged to write as much or as little as they wished (see sample questions at the end of this appendix). These questions explored the subject's thoughts, feelings, and experiences with regard to a wide variety of aspects of online and mobile connecting and interaction. Subjects were asked to consider and probe their own behavior in and responses to portable communities and their impact on people's selves, relationships, and societies. They were asked to respond to a number of issues and most, very generously, did so. In general, subjects were extremely forthcoming, and although interviews like this can not be "timed," the overwhelming majority of interview subjects contributed lengthy, multipage, single-spaced responses.

There are several reasons why people agree to take part in research such as this. In this case, I believe many of my interviewees felt the topic was intrinsically valuable and relevant (to their lives, certainly) and worth their time and attention. In addition, in this and my prior large project I have sensed unusual interest, curiosity, and even excitement in being given the opportunity to talk about (and be heard on) a topic that is rarely discussed. It gives the individual, in effect, "permission" to "vent," in a rather cathartic way, regarding something that is rarely discussed yet may be somewhat central to their lives.[13] As members of portable communities, many of them share with me an interest in the dynamics of such groups. The result was 87 very fruitful interviews that taught me much about the experience of portable communities and shaped my analysis, and indeed the whole project, significantly.

I also asked subjects for the following pieces of demographic information: gender, race, age, and occupation. I did this simply to provide a general overview and illustration of the sample; generalizability to a larger population is not an aim of this research, and, furthermore, the researcher has no control over the subjects to whom she is referred. Most of the respondents were Americans, although 3 of the 87 either hail from or live in other countries. The sample had a gender mix of 70 percent female and 26.5 percent male; 3.5 percent did not reveal gender. Its racial distribution is 75 percent white, with the remaining subjects distributed rather evenly among Hispanic, Asian, African American, and other race/ethnic designations (see appendix 2; subjects were free to respond as they wished with regard to race and ethnicity; I report these designations as they did). Agewise, the sample spanned ages 18 to 67 and skewed toward the under-30 demographic, with 33 percent aged 18–21, 34 percent aged 22–29, 8 percent aged 30–39, 3.5 percent aged 40–49, 15 percent aged 50–59, and 2.5 percent over 60. The median age of the sample was 23. Finally, a number of very different occupations was represented, with "college student," "teacher," and "manager" the most common; it is primarily a middle-class to upper-middle-class sample, though at least five of the college students interviewed reported coming from lower-working-class homes (though I did not ask them specifically for this information; again, see appendix 2 for more detailed profiles of the interviewees in which subjects' backgrounds, interests, and demographic details are provided).

When all the interviews were completed, my undergraduate research assistant Laura Napolitano (a senior sociology major at the College of Saint Elizabeth at the time of this research) and I coded them for the absence and presence of the criteria I initially considered most critical: emotion, intimacy, fun, playfulness, sociability, and practicality. Over time, additional dynamics and activities were included and coded as subjects told me of the pertinence of such practices: these include finding old friends online, the microcoordination of activities, the movement of relationships from online to offline, and the use of portable technologies in emergencies and for self-expression. After a time I worked inductively, drawing out additional major and minor themes as they emerged. Other issues that became apparent only in the course of the interviewing included the use of technology compulsively, to ameliorate boredom, to procrastinate, and to create online content like personal profiles and blogs.

The identification of these issues sometimes required an additional interview session with people previously interviewed. From time to time I also devised new questions that were incorporated into interviews with those still to come. That is why there is no one hard and fast script from which I worked; the interviews were reworked and updated frequently and follow-up interview sessions, sometimes occurring months later, were common. From the data collected within the coded categories and the emergent themes, excerpts of the interviews that best illustrate relevant concepts were selected; those are the ones included in the book's text. A quote from each individual describing his or her experience in portable communities is also found in each of the profiles in appendix 2.

In excerpting these interviews, I protect the confidentiality of subjects by assigning each a pseudonym in the form of a username that corresponds to some aspect of their online behavior. I also report all information with regard to these individuals generally enough that their identities are certain to remain fully protected, reporting occupations generally enough to ensure their privacy, altering nonessential details of the subject's job or stated interests if doing so seemed necessary to maintain confidentiality. In addition, I protect the privacy of all the groups, communities, and sites mentioned and cited by the subjects, only describing them generally, and altering nonessential details that might provide clues to the group's identity.[13] I do not connect any individuals, even though they appear here anonymously, to any specific groups or sites. In some cases I edited subjects' responses very lightly, either for clarity or to further preserve their anonymity and that of the groups they mentioned.

I followed the literature on the recommended strategies and procedures for successful electronic interviewing in designing, conducting, and presenting these excerpts.[14] As the interviews were semi-structured, each interview proceeded in its own way. I therefore provide the following consent form and questions only as a guide and general illustration of the interview process, rather than as a script.

STATEMENT OF INFORMED CONSENT

I am a sociologist and writer on the topic of online life. I teach and conduct my research at the College of Saint Elizabeth and am currently studying the nature of the experience of being a part of online and mobile groups . . . its appeal, its drawbacks, and its overall impact on you and your life. I would really like to include the input of people like yourself (as long as you are an adult, 18 or older) to help me understand and write about it accurately.

The information I gather may be used and quoted in a published book or scholarly article ("Portable Communities: The Social Dynamics of Online and Mobile Electronic Connectedness"), but all identifying characteristics of respondents would be changed, and no usernames would be mentioned, so your privacy would be fully protected. And the online groups you are a part of would only be described very generally; their names and locations would not be given—to protect their privacy, too. If you choose to respond, it is understood that you consent to this and are an adult, age 18 or older. If you would like details as to how the project is progressing and when it may be published, you may email me at any time in the next 12 months and I will gladly provide you with an update, or you may also contact the chairperson of the college's Institutional Research Board (which has approved this research) at . . . [some personal information excised].

So if you would like to tell me about your experience as part of online and mobile groups, please respond to the following questions and the demographic request (a through d) that follows, in a return email to me by [date], writing as much or as little as you like in response to each part of each question. If you prefer, you may print out your responses and mail them to me at Dr. Mary Chayko, Department of Sociology, College of St. Elizabeth, 2 Convent Road, Morristown, NJ 07960—and you need not attach any identification to your response.

Thank you! You will be assisting my research greatly and I would be grateful for the opportunity to learn from you.

Sample questions:

• Do you feel you are "a part of" any online groups? Consider message boards, chat rooms, mobile groups, email listservs, websites, and blogs you may visit regularly, classes, games, MUDs, or something else. Please describe these groups.

• Which of these groups do you visit most often? List as many as you like. You may use their exact names or web addresses, or just give a description of the group—and please also note the type of group it is. (Is it a message board, chat room, blog, or whatever?) Then next to each item on the list, please answer the following: A. Approximately how long have you been visiting it (in months or years)? B. Do you visit often (1–2 times per week), occasionally (1–3 times per month), or rarely (less than once a month)? C. When you visit, do you usually interact with someone, or just lurk?

- For each group you visit with some regularity, please tell me a little bit about the experience of participating in it. What does participating in each group "do" for you? Why do you visit? How does it feel to be a part of the group?

- How in general has participation in online groups impacted your life? Try to think of positive and negative aspects. Has it affected your identity, your relationships online or offline, your social life?

- How do you balance your online life with the rest of your life? Do you think you are successful at this? Does anything "suffer" when you go online—family life, work, other activities you might be doing?

- How do you decide whether you will use the telephone (mobile or landline), the internet, a face-to-face meeting, or even an old-fashioned letter to communicate something with someone? What kinds of information get communicated by you in each of these ways?

- Do you think children are affected differently by using the internet than those of us who are older? How/why? On what do you base this?

- Does going online (or mobile connecting or text messaging) ever boost your mood and make you feel better? If so, how—in what circumstances? Can you give me an example or two? Does it ever make you feel worse—mentally or emotionally or even physically? If so, can you give me an example or two?

- Do you ever connect via computer or cell phone just to alleviate boredom, or to "hang out," or somehow make life just a little more interesting? If so, can you describe why and how, and maybe relate an example or two?

- If you "IM," text message, or read or write a blog regularly, could you comment on the appeal of such activities? What do they "do" for you? Are groups or communities created as you do so? What are the satisfactions, if any? Are there any drawbacks?

- Do you carry your cell phone—or a handheld PDA (personal data assistant) like a Blackberry, or even a portable computer—with you all (or almost all) the time? Could you comment on what it feels like to be so very "in touch with" or "available to" others, as we so often are with the internet and cell phones? How does it affect you?—think about the upside and the downside—and how are you affected when you are "out of touch"?

Request for Demographics Information and Additional Interview Subjects

I would appreciate it if you could also please note the following, for demographic purposes only: (a) your gender, (b) your racial or ethnic background, (c) your age, (d) your occupation—state only in general terms (i.e., a teacher, a salesperson, student, etc.).

Finally, if you think that I would gather useful responses from members of any online groups that you are part of, would you let me know and pass along the group's URL (web address)? Similarly, if you know any adults (18 and older) who are part of portable communities and whom you think would give me thoughtful responses to the above questions, please pass along their emails to me so I may consider including them in my sample.

Appendix 2

PROFILES OF INTERVIEW SUBJECTS

Activist, Hispanic female, age 28. This accountant spends about an hour a day on two electronic mailing lists devoted to political change and minority empowerment. She also spends about a half-hour a day on news and information websites and an hour a day taking online classes and participating in groups associated with the class. She surfs social networking sites and uses her handheld smart phone to access email and make calls. "I do carry my blackberry all the time. It can sometimes be overwhelming because normally I wouldn't have access to my clients and the office until I was back in the office, and now I can be reached anytime anywhere."

AnimeLover, race not provided, female, age 20. AnimeLover enjoys websites that have to do with Japanese animation. She has been an active, involved member of seven such sites for the last few years. In them, anime is presented, created, discussed, and reviewed; one of them is like a blog. She visits at least one of them daily, sometimes for hours. AnimeLover works in a tutoring lab, uses email regularly, and frequently, her cell phone. "I don't get out as much as I could when I'm online. I also don't have too much to talk about with other people not into anime. And another thing is that it is easy to gain weight while sitting online for hours. You just don't notice until it is too late."

Blogger1, Caucasian female, age 19. A college student, this individual enjoys reading and writing blogs and visits two social networking spaces several times a month to facilitate this. She also IMs and emails often, and carries a cell phone with her at all times. "When I am bored I will go online to see if my friends are on or to find something to do but this is rare. I am not one of those people who 'surf the web' for interesting things to do. I'd rather be out with friends."

215

Blogger2, Caucasian female, age 22. Blogger2 has been part of a website that facilitates blogging for five years, visiting four to five times a week and updating her own blog once or twice a week. She also blogs on a separate social networking site from time to time and every so often checks out the discussion boards that feature news on her favorite musical artists. "Online groups have positively impacted me because without them I'd have lost touch with many of my friends that don't live close to me. . . . These are people I never could have called on the phone to keep in touch with."

Blogger3, Caucasian female, age 21. This salesperson writes a blog once or twice a week and regularly visits three social networking sites: one two or three times a day for almost a year, one related to families three or four times a week, and one much less often. She also uses email frequently. "If I can think of even a minor excuse, I'm logged on!"

BookLover1, Caucasian female, age 20. College student BookLover1 has been visiting a fan site for her favorite book series biweekly for three years. She also stays in touch with groups of friends on two social networking sites that she has been visiting daily for several months, text messages frequently, and is never without her mobile phone. "I often go online or call people when I'm bored just as something to do. Usually around homework time . . ."

BookLover2, Caucasian female, age 23. This college administrator is part of a "web group" devoted to the appreciation of novels in the horror genre. She visits its bulletin board two or three times a month. Every other day or so she also logs onto a social networking site. She writes a blog once a month or so and uses her mobile phone fairly often make calls, send quick messages, and send photos to family in other states. "I think being connected to others is important. I am able to stay in contact with friends and colleagues that have moved across the country."

Chatrooms1, African American female, age 21. Management trainee Chatrooms1 can at times sit in social and medical chat rooms "all day." She IMs her friends and family almost daily and has also taken classes online. "It doesn't matter what time it is, someone is always online."

Chatrooms2, Caucasian male, age 51. This scientist has taken online classes and also enjoys online chats related to some of his more uncommon interests and hobbies. He takes his laptop with him whenever he travels and uses it then to retain a feeling of connectedness to his job, friends, and family. He checks email frequently even though he doesn't receive much.

"I have to admit that I get a little frustrated when I cannot connect. On my wishlist is a PDA that can connect to the internet."

E-dieter, Caucasian female, age 28. Besides using a website and online weight loss program daily, this college administrator uses an online course management system a few times a week to communicate with students and help develop online courses. "I have yet to venture into the world of social message boards or online chat rooms. I find them impersonal and a bit scary."

Gamer1, Caucasian male, age 18. This college student is involved mostly in online multiplayer games and associated message boards, visiting them at least every day or so, posting fairly often and even helping to build a wiki. "Do I waste a lot of time on the computer? Definitely. Do bulletin boards aid in this theft of time? Yes. Has anything been seriously impacted as a result of this? No."

Gamer2, Caucasian female, age 21. This student visits several sites where games are played three or more times a week per site, and also participates in an electronic mailing list consisting of all those with whom she once went to summer camp. "I enjoy being online but I can step away to spend time with my family and friends. I am pretty much successful but there are times that I get caught up in what I am doing online and don't pay attention to the time and to what is going on around me. So this can hurt me at times."

InfoGathering1, Hispanic (Puerto Rican) female, age 22. A customer service representative, InfoGathering1 visits a general website daily to obtain news, information, and gossip. She has visited this site for approximately six years. She is also in regular email and IM contact with face-to-face and "online friends." "When I am online my boyfriend feels like I am excluding him from my online time and that I am doing something without him. That is probably the only time that makes it difficult to really sit and participate in online groups."

InfoGathering2, African American female, age 21. College student Info-Gathering2 has been a member of numerous portable communities: several chat rooms and message boards on topics of interest to her that she visited everyday until she decided to cut down on her time spent online. She has also taken online classes and uses email and IM frequently. "Every time I go on, I have to check to see who's on."

InfoGathering3, Hispanic female, age 21. This student considers herself part of online groups that are organized around general news and information, shopping, and school. She visits six or so such sites occasionally, and has done so for approximately a year. She uses a mobile phone fairly constantly after 9 p.m., when minutes are free, to stay in contact with her groups of friends, and uses email as well. "I am a person that does not go online very often. It does not affect other things that are going on in my life. The only thing I can say is that sometimes it does come between me and my homework."

InfoGathering4, Asian female, age 29. InfoGathering4 is a member of more portable communities than she can name, but primarily those for obtaining news and information and job-related tips. A technology consultant, she contributes to most of these communities three to four times a week. She has also taken an online class, emails, and for six months kept a blog in which she had written every three days or so. "I was too frequently visiting the news sites only to realize I was getting the redundant information on each site."

InfoGathering5, African (Egyptian) female, age 19. InfoGathering5 enjoys visiting news and information groups everyday that "help me exercise my mind" and, twice a week, a group that helps students find scholarships and jobs. She also writes a blog occasionally and uses emails and cell phone to stay in touch with groups. "Participating in online groups has impacted my life in a positive light. I balance my online life quite well with my other activities in my life especially school work."

InfoGathering6, Caucasian male, age 19. College student InfoGathering6 is actively involved with a few online groups dedicated to civil liberties and free speech, visiting from one to five times a week, and responding, performing research, and obtaining information on the sites often. "The participation in these groups has been beneficial. It has shown that there are people out there with different outlooks on life than mine. At the same time it has caused me aggravation because some people are incompetent and add irrational thoughts to the conversation or topic."

InfoGathering7, Caucasian male describing himself as Jewish, age 54. This doctor's portable communities are the clients and family with whom he keeps in touch, primarily by cell phone. He browses some work-oriented message boards and websites for information and research, but otherwise is not a major participant in online groups. "I have never liked things such as chat rooms as I think they are impersonal and I am very cautious."

InstantMessaging1, Caucasian female, age 21. InstantMessaging1 has IMed on a daily basis for over four years. A college student, she can IM with five or so people at once and can spend the whole day on the computer. "I think that I sometimes find myself being rude to people who are trying to interact with me face-to-face because I am too busy on the internet."

InstantMessaging2, Caucasian female, age 22. This graduate student has been part of the Instant Messenger community for several years, uses email daily, and also visits special interest websites. For much of that time, she has been "online 90% of the time; on a normal day I would visit my computer to talk or check other members' away messages . . . it took up too much of my time, however. I can imagine that the time I spent 'checking away messages' or 'playing the away message game' could have been spent doing something else alone or with others."

InstantMessaging3, biracial (Irish and Trinidadian) female, age 22. This teacher uses Instant Messenger weekly, using it to keep in touch with her buddy list and friends both near and far. She used it more frequently in college, but much less often since graduation. "If I do not have anything to do, which happens, some people will pick up a book and read, watch a movie, exercise or go out, and others, including myself, will go online."

InstantMessaging4, Caucasian female, age 22. InstantMessaging4 considers herself part of an IM community, participating very frequently for four to five years. She also considers herself part of an email group that contacts one another often, and uses her cell phone often as well (especially after 9 p.m. or on weekends, when calls are free). "It is hard sometimes not being around a computer because you are wondering what your friends are doing or wondering if someone you haven't spoken to in a while has IMed you."

InstantMessaging5, mixed race ethnicity (Spanish and West Indian), age 20. Instant and text messaging are the technologies of choice for this cashier. She uses them most days, along with the occasional email, to stay in contact with her groups of friends. "I IM because most of the time it is cheaper and easier than calling someone and if someone is not available I can leave them a message."

InstantMessaging6, mixed race, age 22. This teacher prefers to use IM to stay in contact with groups of friends and family. She uses it daily and uses email often as well, sometimes to set up dates to get together with people. "For me, this allows me to keep in touch with people I miss but cannot see often."

InstantMessaging7, Caucasian female, age 20. InstantMessaging7 is a college student who is part of three social networking sites that she has been involved with for the last six months to a year and a half. She visits at least one every day and usually lurks, although she used to keep a blog on one and once wrote in it fairly often. She is also fond of using instant messaging and her cell phone to contact and catch up with people. "My blog was really random and scatterbrained. I tried to steer away from any personal matter that I did not want to be accessed. I have a real journal for that."

Lurker1, Caucasian female, age 20. College student Lurker1 has visited two social networking sites regularly for about a year and has instant messaged for many years. She visits both sites every day, but lurks much more often than she posts. She also emails and text messages frequently and always has her mobile phone with her. "My participation in online groups has given me the opportunity to stay in touch with others and have some fun."

Lurker2, Caucasian female, age 22. Two social networking sites are the portable communities of choice for this elementary school teacher. Lurker2 has been a "mostly lurking" member of them for five months, has read several blogs and written in one, and also uses email fairly frequently. "I have started speaking to people online that I haven't contacted in years."

Lurker3, Caucasian female, age 19. This student lurks on a social networking site that she visits every day and, occasionally, contributes messages to. She also IMs daily, emails the professors and older people in her life, and stays in touch with friends and her sports team by mobile phone. She has also read blogs, mostly in the past. "I can easily send a friend an IM to let them know I'm thinking about them—rather than call them when I don't really have time to talk and neither do they."

Lurker4, no demographic data given. Lurker4 has an account on a social networking site and likes to look at pictures there, "pictures of famous people or my friends," but "the truth is I am not very fond of [the site]." Lurker3 IMs and emails occasionally, and carries a cell phone all the time. "I feel really good having a cell phone all the time, because this way my parents and friends know where I am and can get in contact with me if they need me."

MessageBoards1, Caucasian female, age 41. This educational consultant primarily visits online education-related message/discussion boards to complete or gain information for specific tasks, and has done so erratically—sometimes often, sometimes rarely, depending on her work schedule, for the past two years. "You can be short and to the point on the internet without being rude."

MessageBoards2, no demographic data given. MessageBoards2 visits several message boards, a social networking site, and an email circle at least one to two times per week. IM and email are also favored activities. "They help me keep in contact with a lot of my friends and family that are out of state and all over the world."

MobileUser1, Caucasian female, age 55. MobileUser1 belongs to a couple of online singles groups that she has visited occasionally and that have resulted in three or four face-to-face dates. This social worker also IMs her friends and joins online chats, writes the occasional old-fashioned letter, and carries her cell phone "everywhere, even to the bathroom. I feel closer to my family and significant other due to constant communication on a daily basis. I must speak to my boyfriend at least 4–5 times a day and my kids usually once a day."

MobileUser2, Caucasian female, age 22. Student MobileUser2 always has her mobile phone with her, often text messages, and visits a popular social networking site several times a day. She sends messages and lurks, and uses email as well. "Being part of this group is just fun. Many of my friends and family are members and I choose to be a part of it."

MobileUser3, Caucasian female, age 45. Librarian MobileUser3 belongs to an electronic mailing list dedicated to home organization, which she has visited for two years but to which she does not post, is taking an online class, and has participated in education-related emailing, often to a group, daily for the past year. She checks her children's social networking site accounts a few times a month, and text messages, IMs, and emails frequently. She also takes her cell phone with her everywhere she goes. "I am a bit of a control freak and having these devices has made the problem worse."

MobileUser4, Caucasian female, age 20. MobileUser4 has been using email, chat rooms, and IM to network people together for the better part of five years. She also uses mobile phones for talking and text messaging frequently. "In some situations, like a quiet office where you don't want to talk out loud to someone on the phone, text messaging becomes quite convenient."

MobileUser5, Caucasian female, age 20. This college student mostly finds her portable communities on a social networking space, which she visits almost every day, reading and writing messages and blogs. She IMs with her best friend every day, emails, and takes her cell phone with her everywhere. "Going online and mobile connecting gives me a 'pick me up.' Talking to a friend, getting advice, helping someone out, or getting directions are all great."

MobileUser6, Caucasian male, age 28. The primary portable community of MobileUser6 is a popular social networking site that he visits often. This project manager also text messages groups of friends "constantly" and his mobile phone "never leaves my side." Participation in online groups "has impacted it in a positive way by expanding my circle of friends to those people I don't normally have contact with due to geographical location or conflicting schedules."

MobileUser7, Caucasian male, age 27. Salesperson MobileUser7 is on an alumni electronic mailing list and a work-oriented list that helps him network with business contacts and colleagues. He also emails and IMs, does a lot of work and research online, and usually carries his mobile phone with him. "I am not dependent on having this technology available at all times but it is a luxury. When I am out of range and I can not get my cell phone/internet access I am reminded of the luxury."

MobileUser8, Trinidadian male, age 25. This corporate manager is part of many online groups, from a variety of websites to games, which he accesses every day via PDA (Blackberry). He also IMs frequently. "I feel compelled to check my phone all the time for an email or something like that. I want to keep in touch with people even when I am not at work. In some ways it promotes micro-managing but then again it fills in for the times that we cannot interact face-to-face with one another."

MostlyEmail1, Caucasian male, age 26. This graduate student initially did not consider himself to be a member of any portable communities. As someone who maintains a professional network of colleagues and their organizations online, though, he fit the criteria of the study, with email his primary means of contacting them. "Generally, I'm not a person with a million friends, so my face-to-face/phone contingency is pretty small. I use email to maintain my formal networks."

MostlyEmail2, Caucasian female, age 22. This elementary school teacher also did not initially consider herself a member of any online or mobile groups. But during the course of the interview she realized that she was keeping in touch with her community of college friends by fairly regularly emailing and IMing them. She also carries her cell phone with her at all times so that friends and family can always reach her. "Online always comes second," she says. "I always put my work and my family first."

MostlyEmail3, Caucasian female, age 35. A college administrator, Mostly-Email3 has developed a curiosity about dating and friendship social networking sites, participating at a minimal to moderate level, never very frequently. She relies on email to maintain connectedness with people, checking it compulsively, and her cell phone, on which she prefers to receive messages and then call people back at her convenience. "There is something about the delayed response that I feel safer with, or prefer."

MostlyEmail4, Caucasian female, age 38. MostlyEmail4 is part of several mailing list communities, mostly revolving around college friends and activities, that she takes part in once or twice a week. She is a hairdresser who feels a sense of pride to be part of her communities, but notes with regret the activities that have fallen by the wayside: "I am not participating in other things because I am too busy on the computer. Social life, crafts, and hobbies may suffer, while family and work take priority."

MovieBuff, Caucasian female, age 20. This college student has belonged to three social networking sites for the past one to two years, to stay in contact with friends and to post reviews of movies and read others' reviews. She reads blogs once a week or so and uses a cell phone for talking and texting. "[My social networking space] is sometimes the only communication I have with people I haven't seen in a while."

MusicLover1, Caucasian female, age 28. MusicLover1 enjoys rock music and visits a message board for fans of her favorite musician every day, though she only posts on the board occasionally. She is fluent in several languages and works as a translator. "I have a tendency to occasionally check the board at work so this might hinder my productivity. Otherwise, I find that I don't let the message board take me away from other aspects of my life."

MusicLover2, Hispanic female, age 21. MusicLover2 takes part in several online groups related to reggae music in the form of message boards and social networking sites. She has been visiting some of them for four years, and visits at least one each day; she used to visit them five times a day until she took a position as an accountant. She also emails and text messages people with great frequency because "with mobile connecting you keep in touch with your friends even if you are somewhere and you can't call them."

MusicLover3, Caucasian female, age 21. This salesperson is part of "many" online groups related to her favorite musical groups and genres, visits them daily, and interacts actively on all of them. She also writes in two blogs regularly, one on a social networking site, and reads blogs as well. "People can comment on things and help you out and you can do the same for others. I love it."

OnlineInstructor, Caucasian male, age 52. This college professor has been part of an electronic mailing list for college alumni for four years. He contributes regularly and often (one to two times a week) to the discussion. He has also taught online classes for three years and has a number of friends that he interacts with on a weekly basis via email and mobile phone. "These online groups have greatly increased my ability to stay in touch with my friends and associates, and I can provide instruction to students that simply could not take classes on campus."

OnlineStudent1, Caucasian female, age 50. OnlineStudent1 has spent most of her internet time in the last four years in online classrooms and related discussion boards and work groups. A medical professional, she also plays chess online a few times a week, uses email frequently, and carries a cell phone most of the time "when I remember and when it is charged. I actually learned to turn off my beeper at the end of the day to remind myself to disconnect from the day."

OnlineStudent2, no demographic information given. OnlineStudent2 seldom goes online except to take online courses, do research, and communicate with others in her classes. She does communicate with friends overseas by IM and works via internet and mobile phone as well. "It alleviated having to attend class on campus thus creating less of a hassle . . . it's also a great way to connect while at work."

OnlineStudent3, Hispanic female, age 26. This teacher and online student primarily considers herself part of two portable communities, an online class on which she spends about two or so hours per week and a social networking site, on which she spends about an hour and a half per day. She reads comments others have left for her, emails and IMs, and adds things to her page. According to OnlineStudent3, "it just kills time."

OnlineStudent4, Caucasian female identifying herself as Jewish, age 39. A salesperson, OnlineStudent4 says that her online class and the discussion boards that are part of it comprise her primary portable community. She does text message and use email on occasion. "I text message when

I do not want or need an immediate response. I hope it is less intrusive than calling."

OnlineStudent5, Caucasian male, age 59. This manager in a marketing firm takes online courses and feels part of those communities, and uses email and IM constantly at work. He has also visited a social networking site to try to locate old friends and keeps his cell phone "always on." "At work we use IM constantly as a means of communication. The appeal is that it is a tool to get information in an environment where folks are very busy and on calls and may not be able to talk directly as needed."

ParentingForum1, Caucasian female identifying herself as "Jewish living in Israel," age 31. ParentingForum1 is part of two portable communities: one is a medical electronic mailing list that she has subscribed to for over seven years and was once much more active on but now mostly scans posts, responding only rarely, and the second is a message board devoted to pregnancy and parenting that also supports chatting and IM-ing. She participates in these functions of the group a few times a day and wholeheartedly, along with emailing and mobile telephony. "I do check my messages regularly, but prefer not to be available every second of the day. If it is really important, I can be reached somehow, otherwise it can wait."

ParentingForum2, Caucasian female, age 28. A computer technician, ParentingForum2 visits numerous websites and blogs a day, and feels connected most to a community of women on a parenting-oriented site where they can email and IM. They also talk to one another on the phone; she makes sure her mobile phone is with her at all times. "We know about one another's personal lives and support each other on a daily basis."

SocialNetworking1, Caucasian female, age 35. This librarian visits many types of social networking sites (including blogs and dating sites) and bulletin boards (pregnancy, medical advice, gaming), generally a few times a week or more, and has for several years. She has also come to rely on text messaging and is "never without my cell phone. I love it. Period."

SocialNetworking2, mixed race (African American, Puerto Rican, Native American) female, age 21. SocialNetworking2 has used instant messaging for up to two hours a day for five years. More recently she has begun visiting large social networking sites regularly, to keep in contact with friends, and text messaging on her mobile device, which she carried with her most of the time. "I just use online groups as an addition . . . never would I allow myself to become so consumed with the online culture that my everyday physical relationships would suffer."

SocialNetworking3, Caucasian female, age 29. A popular social networking site is the portable community of choice for SocialNetworking3. She visits five times a week and has done so for the past year or so. She does not use email or her cell phone very often or with any regularity. Nevertheless, "since joining [a social networking site] I do keep in touch with friends and family more often."

SocialNetworking4, Caucasian female, age 32. SocialNetworking4 is a member of a social networking site and a dating site. While her presence on the networking site and her use of IM has receded, she is fond of email and text messaging and old-fashioned letters. "I mostly check sites and all that at night instead of watching TV. . . . Sometimes friends call me and tell me to go online so that I can chat with them."

SocialNetworking5, Caucasian female, age 20. This student keeps up with groups of people in a social networking site that she has visited every day for the past year. She also emails and uses a cell phone and IM from time to time, mostly to stay in touch with friends and family. "I find myself on the site when I should be studying . . . but I think family relationships are benefiting from online use such as email and instant messaging."

SocialNetworking6, Caucasian female, age 21. This college student's primary portable community is found on a social networking site that she visits every day, ordinarily just lurking and updating her profile. She has been on the site for over a year. Email and her cell phone helps her stay connected to the others in her life, and she does read blogs, though rarely. "It is a huge way to waste time when I could and should be doing something more productive."

SocialNetworking7, Caucasian female, age 20. SocialNetworking7 has been part of three social networking sites for the past year to year and a half. She carries a cell phone with her all the time. She occasionally leaves comments for others, reads their blogs as often as a few times a week and keeps in touch with high school friends attending different colleges. "It can have negative impacts when people that you don't intend to see your blogs read them, but can be easily fixed by making the entries only visible to certain people."

SocialNetworking8, Caucasian male, age 19. This student says that although he is part of several online groups, he identifies most with the one on his social networking site, which he checks three to four times a day. He also uses email and mobile phones to stay in touch. "I guess on [my social networking site] I just change my profile in a flamboyant way to tell everyone who cares what is going on with me at the current time. Seems after thinking about it what a waste of time!"

SocialNetworking9, Hispanic female, age 21. Two social networking sites and an online class comprise most of this student's portable communities. She considers herself a member of one social networking site in particular, having been a member for almost a year and visiting very often, though generally lurking. She also emails people often and carries her cell phone with her everywhere. "You have the opportunity to find people you have not seen in years . . . I have come in contact with people that I graduated eighth grade with."

SocialNetworking10, Caucasian male, age 18. SocialNetworking10 visits two social networking spaces most often to lurk and read messages. He also blogs and leaves comments of his own on other blogs from time to time. He text messages and IMs to stay in contact with and make plans with friends. "It is a good way to keep in touch with friends that I don't get to see or talk to all the time."

SocialNetworking11, Caucasian male, age 19. This college student is part of two social networking sites on which he participates multiple times a day. He is also a member of a large email circle that he participates in frequently, and IMs and uses his cell phone often. "I will say that when I'm talking to someone and find out that that person uses [the same site], I do feel some sort of connection."

SocialNetworking12, Caucasian male, age 23. Firefighter SocialNetworking12 enjoys communicating fairly often on two social networking spaces and a few message boards dealing with automobiles as well. He also relies on his cell phone for both talking and texting multiple times daily, and has weaned himself off instant messaging. "Now I mainly contact people through [my social networking site] or text messaging."

SocialNetworking13, African American male, age 23. The portable community of choice for this graduate student is a popular social networking site. He also emails but does not use most portable technologies, including IM and cell phone. "[The site] is good for staying in touch with people that you no longer live close to, and keeping relationships alive . . . I did reconnect with a couple long lost friends."

SupportGroup1, Caucasian female, age 57. A medical support group that meets via electronic mailing list is this college administrator's primary portable community. She has visited it at least weekly for the past five and a half years, though she usually lurks, and only posts to the group every other month or so, though she may contact a member of the group directly more often. She also emails people regularly and carries a mobile phone with her at all times. "My life is full all the time with work, social engagements, doctors visits, family obligations, home maintenance, etc. So I have to deliberately carve out time each week to keep abreast or otherwise the list becomes an overwhelming mass of mail."

SupportGroup2, Caucasian female, age 61. This retired educator is an active part of two portable communities: an electronic mailing list that she accesses daily and to which she posts about once a week, and a medical chat room that she enters several times a week, usually as a lurker. She also writes a newsletter to her family and friends about once a month that has helped many members of her extended family to reconnect, and sends them emails. "I cannot think of any negative aspects of online life other than carpal tunnel (LOL) and maybe being guilty of spending too much time on my PC. I don't even try to balance my life . . . I have a disease, 61 years old, on SS disability, children all gone, sooooooooo, I indulge myself."

SupportGroup3, Caucasian male, age 55. SupportGroup3 is a manager of a technology firm that has been visiting his medical support group, which takes the form of an electronic mailing list, for seven years. He visits the group at least twice a day most days, is a frequent poster, and also interacts privately with group members via email and in a few cases by phone or face-to-face. He has also used a popular online dating site. "Now, seven years later, I am starting to establish a [face-to-face] social life as best as I can and spend significantly less time on the list."

SupportGroup4, Caucasian male, age 55. Salesperson SupportGroup4 visits his medical electronic mailing list daily, and has done so for the past nine years. He lurks about 95 percent of the time and responds the other 5 percent. He has visited a social networking site on occasion, reads a blog that deals with his disease, and owns a cell phone but rarely uses it. "Children may feel that the internet is the only tool but us 'older' folks know that the internet is one of several tools available as a resource."

SupportGroup5, Caucasian female, age 67. Support Group5 works in real estate and is part of two online electronic mailing lists that deal with a serious disease. She has been a member of both for a year and interacts actively on them every day. She also emails regularly and carries a mobile phone that she uses only for business. "I belong to a monthly book club, lots of friends, pool time, BBQs. But even communication with friends is almost entirely by email in between face time."

SupportGroup6, Caucasian female, age 59. This retired educator is part of two portable support groups, for two different medical problems, each of which take the form of electronic mailing lists. She has been on one for eight years and one for three, visits both daily, and posts to them fairly often. "The internet has been a much better way to communicate. I use it to keep a long list of friends posted on my condition, which sure is a simple way to let people that I don't talk to on a regular basis know how I am doing."

SupportGroup7, Caucasian female, age 48. A medical professional, SupportGroup7 is a new (six-month) member to her portable support group, a disease-oriented electronic mailing list. She usually lurks, though she responds occasionally, and has even emailed members off-list. "I usually only go online when everyone is out or busy and not around me. I don't like them to know I am doing research into the disease."

SupportGroup8, Caucasian male, age 58. This retired salesperson is a member of nine portable communities, some for as long as three years. He ranges on these from lurker to extremely frequent participant, especially on the electronic mailing lists related to his disease, his military experience, and his political and activist activities. "I am online as much as ten hours per day. I make a point to do it when my wife is sleeping or otherwise engaged in other activities as much as possible, although there are a number of hours in the day when she would rather I be with her (we are both retired and totally disabled). I am only partially successful at this."

TVFan1, Caucasian male, age 26. TVFan1 visits his favorite music and television message boards at least every other day, posting more often on the music forum than the TV one, but generally eschewing social networking sites. A computer programmer, he always has the computer on, "from the time I wake up to just before I go to bed. When I'm home, it is only a room away, so I know if I'm bored and nothing is on TV and I have nothing else to do I just go to my computer and I'll always find something—whether it's playing a game, or visiting a message board, or playing poker, or just surfing, there's always something to do. I carry my cell phone with me all the time and to be honest, I hate it."

TVFan 2, Caucasian female, age 32. This secretarial assistant frequents several message boards dedicated to her favorite television programs and a popular social networking space. She visits her favorite groups daily and the others weekly, and has been doing so for about a year. She mostly lurks but posts when she has something she really wants to add to the conversation, and also enjoys meeting friends and discovering new music and bands online. She also uses email frequently and visits travel websites. "I usually don't keep my cell phone on so most people don't even call me on it. But I have to say it is funny when it actually is on and rings out of the blue because it's always a real shock to me that someone is actually trying to track me down."

WorkGroup1, Caucasian female, age 24. WorkGroup1 has been using her company's website to transact business and make work contacts for over two years. Although she is a writer, she does not use the website very often, nor does she use IM or social networking sites, but emails friends, family, and members of her working groups (her colleagues) frequently, and carries her mobile phone with her wherever she goes. "I don't know how I would work or keep up all necessary communication without email, and it allows me to stay close with friends/family who are far away."

WorkGroup2, Caucasian female, age 23. This middle manager's online and mobile use mainly revolves around business groups, email discussions, and threads that circle among clients and colleagues. She also emails and text messages friends and family regularly, constantly carries a mobile phone, and has established a social network profile and page where friends can leave her messages . . . "Peer pressure!" she says. "For me, logging on makes me feel 'connected' to my client groups."

WorkGroup3, Caucasian female, age 24. A manager in a midsized organization, WorkGroup3 is an active member of a work-related electronic mailing list and a lurker on a couple of others related to other interests. She checks email often, carries her cell phone everywhere she goes, and has just signed up for IM and has so far sent exactly one IM. "I am a big fan of checking my email. I look forward to receiving all types of messages but unless I know that an important message is coming I try to leave the work and social related emailing to my work hours."

WorkGroup4, Caucasian female, 23. A job recruiter, WorkGroup4 uses portable communities to create and work daily within working groups. She kept a blog for two years, writing in it twice a week, and is a member of two social networking sites, one for just over a year and one for four months. She participates in these often, from three to five times a week, generally updating her profile and reading messages left for her. She carries her PDA with her at all times and text messages many times a day. "Email does distract me from my work, as you can tell presently, as I am writing this at work."

WorkGroup5, Caucasian male, age 22. This technical consultant has belonged to a portable community associated with work and work issues, with over 100,000 members in it, for just over seven years. He lurked for the first year and has participated more actively after that. He also IMs "all the time" and keeps his laptop and mobile phone close at hand. "When I am at home I usually never log on, I usually spend time with my family and friends."

WorkGroup6, Caucasian male, age 51. An artist who produces and distributes his own work, this individual is a regular participant in a work-related chat room and has visited several social networking sites, mostly for work and finding old friends but also for dating/romantic purposes. He text messages and IMs daily, and takes his mobile phone with him everywhere. "I connect with friends online when I'm bored, either writing emails or instant messaging just to stay hi and keep in touch."

WorkGroup7, Caucasian male, age 58. This teacher is a member of a work-related online association that provides him with frequent and essential business networking. He is also available by cell phone almost all the time. "A little less accessible would be better!"

WrestlingFan, Hispanic female, age 23. WrestlingFan visits a number of websites dedicated to wrestling, games, music, and travel three or four times a week. She reads and writes blogs regularly and participates on social networking spaces related to her interests and to connect with friends and potential colleagues in her field, college administration. She is an active part of a school-related electronic mailing list, emails frequently, and carries a mobile phone with her at all times. "Connecting or having the ability to have a form of connection with me at all times is very valuable to me."

NOTES

CHAPTER 1. THE PORTABILITY OF SOCIAL CONNECTEDNESS

1. The 87 electronic interviews I conducted for this study are excerpted throughout this book. See appendix 1 for a complete description of the methodology. As stated in appendix 1, I assign a pseudonym to each interview subject that describes something of his or her online interests or habits. More detail is found on each interviewee in appendix 2. The quote that opens this chapter comes from the interviewee I call SocialNetworking1.

2. See Chayko (2002).

3. On the large number of social connections typically made by each of us, see Preece (2000:174). On the large number of social connections we tend to make with distant others, see Caughey (1984).

4. The *internet* will not be capitalized in this book, following Joseph Turow's and Steve Jones's endeavor to encourage the treatment of the internet as a technology embedded in everyday life, "the way we refer to television, radio and the telephone," as Jones has argued, in Schwartz (2002).

5. Grossman (2006).

6. See Legge (2005), Eldridge and Grinter (2001), and Hafner (2006).

7. Chayko (2007). See Wellman (1979 and 2001, esp. pp. 227–28), and Wellman et al. (1996).

8. See Simmel (1964:307–16) on the dyad and the triad as a group, and Anderson (1983) on a nation as a community.

9. Durkheim ([1912] 1965):245–47.

10. Chayko (2007, 2002:40).

11. Tönnies ([1887] 1963). See also Etzioni and Etzioni (1999), Fowler (1991), and Preece (2000).

12. See Bell and Newby (1973), and for a more contemporary, symbolic interactionist argument, Fernback (2007).

13. On the concept of social capital, see Bourdieu (1983), Putnam (1995), Portes (1998), Burt (1997), Katz and Rice (2002:199), Lin (1999), and Kavanaugh et al. (2005) for excellent overviews, particularly where cyber-communities are

233

concerned. On group identity, purpose, and commitment, see Preece (2000:179–81), Wallace (1999), Lea and Spears (1991), and Fernback (2007). On social networks, social ties within networks, and deriving social capital and identity from networks, see Granovetter (1974), Wellman (2006), Haythornthwaite (2005), Wellman and Hampton (1999), Calhoun (1998), Warren (1978), Chayko (2007, 2002), and Venkatesh (2003).

14. For examples of this type of physically relocating portable community, see Stoller and Longino (2004) and Gardner (2004). For a look at a technologically mediated portable community, as I will use the term in this book, see Ling (2007).

15. See Anderson (1983).

16. Social networking sites often support blogs and blogging (LiveJournal is a current example) and many of the applications discussed here (like photo, video, and podcast-sharing). For more on blogs and social networking sites, see Lenhart and Madden (2006), Lenhart and Fox (2006), Blood (2002a and 2002b), Huffaker and Calvert (2005), boyd (2006a and 2006b), and many others to be discussed herein. On wikis, see, for example, Lamb (2004). On avatars, see, for example, Viljalmsson and Cassel (1998).

17. Katz and Rice (2002:252–53) demonstrate that general internet use and the use of mobile phones are highly related. See also Miyata et al. (2005:146), Geser (2004), and Kharif (2006).

18. See Chayko (2002), Kendall (2002), Preece (2000:179–81), and Wallace (1999). For an interesting discussion of how mobile phone use can confer "instant membership" into a community, see Wei and Lo (2006).

19. Rheingold provides a fascinating analysis of the phenomena of online, and later mobile, community. See, especially, Rheingold (1993, 2002).

20. See Chayko (2007), Virnoche and Marx (1997), and Feenberg and Bakardjieva (2004).

21. Thomas and Thomas (1928).

22. Term conceived in collaboration with Eviatar Zerubavel. See Zerubavel (1993). See Chayko (2007) for explicit definitions of this concept and related concepts. For a more extended discussion of the sociomental, see Chayko (2002).

23. See Chayko (2002, esp. pp. 39–40).

24. See Chayko (1993) on the creation and use of such a continuum, and Fox (2004) on "degrees of virtualness."

25. As demonstrated by Hamman (2005), Jackson et al. (2004), Kavanaugh and Patterson (2002), Wagner, Pischner, and Haisken-DeNew (2002), Rheingold (2000 and 2002), McKenna, Green, and Gleason (2002), Hampton and Wellman (2002), and Wellman and Hampton (1999) among others.

26. Boase et al. (2006).

27. Kelly (2005).

CHAPTER 2. THINKING IN TANDEM: COGNITIVE CONNECTEDNESS

1. See Simmel on the persistence of social groups (1898).

2. See Chayko (2002:41), Anderson (1983), and Cooley ([1922] 1964).

3. Durkheim ([1912] 1965:251). See also Chayko (2002:80–84) on the importance of symbols to a community.

4. Durkheim ([1912] 1965:252).

5. Ibid., p. 252.

6. See Chayko (2002:89–96).

7. See Mead (1934) and Chayko (2002:42–46).

8. See Chayko (2002:46).

9. Klastrup and Tosca (2004). See Klastrup and Tosca (2004) for an interesting analysis of massive gaming cyberworlds and their joint "elaboration" and change over time. I provide discussion of groups' orientation to similar images in Chayko (2002:43–45). See also O'Brien (1999) and Walther (1996).

10. Baker (2005:35). On bonding with others in the absence of a visual, see Baker (2005), Curtis (1992:66), Walther, Slovacek, and Tidwell (2001), and Chayko (2002:41–47). On the importance of the mental image in the creation of online gaming worlds, see Klastrup and Tosca (2004).

11. See Jackendoff (1994).

12. Sterling (1992:xi–xii) in Jordan (1999:55–56).

13. Sterling (1992:xii) in Jordan (1999:55–56).

14. Jordan (1999:55).

15. For more on mental mapping and social topography, see Downs and Stea (1977), Gould and White (1974), Jackendoff (1994), Lewin (1936), Leach (1976), and Chayko (2002:30–37) for an overview of socimental space in general and mental mapping in particular. See also Chayko (2007).

16. Joshua Meyrowitz's look at the impact of electronic media on social behavior (1984) precedes the internet era, but provides a classic analysis of place and location in modernity and has strongly influenced my theorization of social connectedness with distant, absent others. This quote is from p. 145.

17. Ibid., p. 308.

18. For more on intersubjectivity, see Schutz (1962, esp. pp. 11–12) and Heritage (1984:54–61). On intersubjectivity, the creation of a common stock of knowledge, and cognitive resonance, see Chayko (2002:25–30 and 66–73).

19. Clark and Brennan (1993:22). Jenny Preece provides a good overview of how common ground theory can be used as a framework for determining the extent to which people understand one another (2000:156–65).

20. Chayko (2002:59). On complementary differentiation in general and its relation to cognitive resonance, see Chayko (2002:59–64).

21. All of these quotes are from Baker (2005:45).

22. See Schutz (1951) and Schramm (1954). Chayko provides an extensive overview of these concepts (2002:66–68.)

23. See Preece (2000:154).

24. See McKenna et al. (2002:14).

25. In McGowan (2004). On the production of dopamine and norepinephrine in romantic attachment, see Fisher (2004) and Psychology Today staff (2004).

26. On the interlinking and synchronization of brains, see Goleman (2006). For more on social synchronicity and entrainment, see Goleman (2006), Strogatz (2003), and Barabási (2003).

27. For more on this, see Chayko (2002).

28. Chayko (2002:67). For examples of this, see Cerulo (1995), Cerulo and Ruane (1996), Zerubavel (1981), Lombard and Ditton (1997), and Dayan and Katz (1992).

29. See Zerubavel (1981).

30. Cerulo (1995:13).

31. Barlow et al. (1995:36, 40).

32. Chayko (2002:3–6).

33. See Fox (2001).

34. Zerubavel (2003:13).

35. Ibid., p. 13.

36. Poster (1995) and Pavlik (2000) make a similar argument; see Thoreau (2006).

37. See Tilly (2006), Mitra (2005), Christian (2005), and Gladwell (2006).

38. Lenhart and Fox (2006). Statistics reported in this book should be considered accurate only as of the writing of this book, which took place during 2006 and 2007 and was completed in late 2007; they are presented here not as definitive but to outline the general scope of a phenomenon. It should furthermore be kept in mind that statistics generated by the Pew Internet and American Life research center were gathered in the United States and should be generalized to U.S. populations.

39. Ibid.

40. Gladwell (2006:80).

41. Okabe (2004).

42. Tilly (2006).

43. See Schrock, Holden, and Reid (2004) and Mason-Schrock (1996).

44. See Etzioni and Etzioni (1999), Gasson (2005), Cook and Brown (1999), and Smircich and Morgan (1982).

45. As argued in Etzioni and Etzioni (1999). See also Flanagin and Metzger (2001).

46. See Tanner (2001).

47. Burnett (2002).

48. Silver calls for this type of research as well in a more detailed call to action (2003:288–90); see also Silver and Massaneri (2006).

49. Such as communities under threat of genocide; see Carmichael (2003).

50. See Tanner (2001) and Mikula (2003).

51. Simmel (1898) puts forth the classic formulation of this.

52. See Jones (1999), Roberts and Videl (1999), and Foot, Warnick, and Schneider (2005).

53. As detailed in Urbina (2006).

54. Siegl and Foot (2004). See Cerulo and Ruane (1996) for a fascinating look at how technology can, in effect, "extend" one's life and bring the dead into a "lifespace."

55. Linenthal (2001).

56. Foot, Warnick, and Schneider (2005).

57. Short, Williams, and Christie (1976)

58. Richardson and Swan (2003:3). See Lombard and Ditton (1997) for a definitive overview of social presence, related research, its most common types, and its consequences.

59. Licoppe (2004:65).

60. Haythornthwaite (2002:182).

61. Burnett (2002).

62. Ibid.

63. See Wei and Lo (2006) and Ling (2004a).

64. See Ito and Okabe (2005:264–66), Gray et al. (2003), and Quan-Haase and Wellman (2002).

65. Nakajima, Himeno, and Yoshii, quoted in Ito and Okabe (2005:264).

66. See Licoppe (2004) and Christensen (2004).

67. See, for example, Loftus (1996), Neimark (1995), Hannigan and Reinitz (2001), and Chayko (2002:112–14).

68. For detailed descriptions of this, see Berger and Luckmann (1967) on the "reality of everyday life" and Habermas (1989) on the "lifeworld."

69. A further articulation of this concept and these social worlds can be found in Schutz (1973), Davis (1983), Berger and Luckmann (1967), James ([1893] 1983), and Caughey (1984). See also Juul (2005) on social worlds that emerge in gaming and Shibutani (1955) on social worlds in mass media use.

70. See Berger and Luckmann (1967), Chayko (1993), Zerubavel (1991), and Goffman (1974) on the "framing" of social experience.

71. Davis (1983:10).

72. See Goleman (2006).

73. See Winnicott (1971).

74. See Chayko (2002).

CHAPTER 3. FEELING CONNECTED: EMOTIONALITY AND INTIMACY

1. As noted in Tanner (2001).

2. See Hardey (2002).

3. For more on each of these properties of connections, see Chayko (2002:101–26).

4. Noted in Else (2007).

5. See Karbo (2006) and Fehr (2005).

6. boyd (2006b). In her excellent overview of the topic, boyd notes that not all social networking sites work similarly; some do not require reciprocity in "friending." See also boyd (2006a).

7. As noted in Karbo (2006:95). See also Cerulo, Ruane, and Chayko (1992).

8. See, for example, Chayko (2002), Wellman et al. (2006), Wellman et al. (1996), Wellman and Hampton (1999), Haythornthwaite (2005 and 2002), and Cerulo et al. (2002).

9. See, for example, Nowak, Watt, and Walther (2005); Walther (1996); Hian et al. (2004), and Hu et al. (2004).

10. Baker (2005:83).

11. Kendall (2002:185).

12. Giddens (1992, esp. p. 130).

13. See Wellman and Gulia (1999:178–81), Cerulo and Ruane (1998), and Reid (1995).

14. In Hian et al. (2004) see Burgoon and Hale (1987) and Prager (1995). See also Chayko (2002).

15. See McKenna et al. (2002:14) for the complete study; see also Bargh, McKenna, and Fitzsimmons (2002) and Bargh (2002).

16. Newsweek Online (2002).

17. Baker (2005:35).

18. See, for example, Thurlow, Lengel, and Tomic (2004:52–53), Walther and D'Addario (2001), Lea and Spears (1992), Baker (2005), Preece (2002), Katz and Sugiyama (2006), and Ling (2004a).

19. Thurlow et al. (2004:53).

20. Noted in Thurlow et al. (2004:53); see Thurlow et al. (2004:52–54).

21. Bargh et al. (2002); see also Murray, Holmes, and Griffin (1996), Walther (1996), and Hian et al. (2004).

22. See Bargh et al. (2002).

23. Walther (1996); see Thurlow et al. (2004:53).

24. As noted by Gavin in ScienceDaily.com (2005).

25. As argued by Thurlow et al. (2004:53) and Suler (2004).

26. Suler (2004:321).

27. Gergen, Gergen, and Barton (1973) as discussed in McKenna (2002:23).

28. As noted in McKenna et al. (2002:10); see Galal (2003).

29. Newsweek Online (2002).

30. ScienceDaily.com (2005).

31. Examples of this are noted in Ling (2004a:162–63).

32. See McKenna et al. (2002:10), Galal (2003), and Hardey (2002:575).

33. As demonstrated in McKenna et al. (2002:11) and Galal (2003).

34. Many examples of this can be found in Hardey (2002:580), Galal (2003), Baker (2005), Sveningsson (2002), Kendall (2002), and Hu et al. (2004).

35. According to one of the subjects surveyed by Baker (2005:125).

36. Baker (2005:127).

37. As noted by several subjects in Kendall's study of an interactive, text-only multi-user domain (MUD) (2002:159–67).

38. See Walther and Parks (2002:76) on this kind of online-offline migration, and Kayahara and Wellman (2007) on how people go back and forth among online and offline sources to obtain cultural information.

39. See Igarashi, Takai, and Yoshida (2005).

40. See Geser (2004).

41. Gross and Simmons (2002:533), paraphrasing Giddens (1994:89–90). For more on our belief that corporations or governments will protect us, even if we believe that they will disclose information about us, see Turow and Hennessy (2007).

42. See Henderson and Gilding (2004). Metzger (2007) finds that consumers are more likely to disclose personal information online if they trust the retailer's reputation.

43. See Hossain and Wigand (2004), Geser (2004), Licoppe and Heurtin (2002), and Hardey (2002). On our trust of Google and its ranking procedures, see Pan et al. (2007).

44. Hardey (2002).

45. See Rosson (1999:80), on the disclosure of personal information anonymously online, Henderson and Gilding (2004) on familiarity and reciprocal disclosure online, and McKenna et al. (2002:16) and Baker (2005) on the development of trust leading to offline interaction. On the development of trust in the absence of face-to-face interaction, see Griffiths (2001:6) in Galal (2003), and Suler (2004). On trust in mobile phone use, see Campbell and Kelley (2006), Geser (2004a), and Castells et al. (2004).

46. See Haythornthwaite (2005:182–83).

47. See Drentea and Moren-Cross (2005).

48. See Preece (1998, 1999, 2000). On reciprocity in social bonding at a distance, see Chayko (2002).

49. See Finn (1999) and Burnett and Buerkle (2004).

50. Radin (2006:591).

51. Ibid. (2006).

52. Ibid. (2006); see also Radin (2003).

53. See Broom (2005:83).

54. Beder (2005). See also Pecchioni and Sparks (2007).

55. See LaRose, Eastin, and Gregg (2001).

56. See Green et al. (2005), Galal (2003), and Lindsay et al. (2007).

57. On computer-mediated communication in general, see LaRose et al. (2001). On the sharing of written emotional disclosures, see Radcliffe et al. (2007).

58. For examples of this, see Preece (1998, 1999, 2000), Ling (2005a), Baker (2001), and Lee (2005).

59. See Preece (2000:83).

60. See Baker (2001) and Preece on "netiquette." (2000:99–101)

61. See Lee (2005).

62. Turkle in Else (2007:10).

CHAPTER 4. PLAYING AROUND: FUN, GAMES, AND HANGING OUT

1. See Fallows (2006).

2. See, for example, Wasko and Faraj (2000), Ridings and Gefen (2004), Sandvig (2006), Glasser (1982 and 2000), and Stephenson (1964 and 1967).

3. Glasser (1982 and 2000) and Huizinga ([1938] 1950); see Sandvig (2006).

4. Zimmerman (2004) and Sandvig (2006).

5. Sandvig (2006); see also Gadamer (1989).

6. See Mead (1934).

7. Huizinga ([1950] 1938); see Suellentrop (2007:61).

8. Noted in Wilson and Tan (2005:394).

9. See Bateson (1972), Goffman (1974), Handelman (1976), in Danet (1997).

10. Danet (1997).

11. In Danet (1997); see also her classic study of cyberplay (2001).

12. See Khazen (2006).

13. See Huizinga ([1950] 1938), Juul (2005:29–34), and Zimmerman (2004).

14. Sandvig (2006) provides several examples of how children do this.

15. Danet (1997).

16. See Ivory (2006).

17. See Jones (2003).

18. McKenna (2003). On women and text-based gaming, see Danet (1996) and Mulcahy (1997). On the disguising of gender online in gaming, see Jones (2003) and Bodmer 2001, and for more on all of this, Thurlow et al. (2004:131).

19. See Yee (2006).

20. Zimmerman (2004).

21. On games as communities, see Huizinga ([1938] 1950:3) and Juul (2005:92). On the content and impact of MMOs and MMORPGs see Khazan (2006), Steinkuehler and Williams (2006), and Fung (2006), and on intensity of involvement in them, see Klastrup and Tosca (2004), Farrar, Krcmar, and Nowak (2005), and Khazan (2006). On the creation of MUDs, see Kendall (2002).

22. Fung (2006).

23. As noted by Utz (2000) and Ridings and Gefen (2004).

24. In Khazen (2006).

25. See Jones (2003).

26. See Juul (2004).

27. See Juul (2005 and 2003).

28. See Johnson (2005 and 2006).

29. Johnson (2005).

30. As discussed in Suellentrop (2007); see Beck and Wade (2004).

31. Suellentrop (2007:62), paraphrasing the theory of game designer Will Wright.

32. See Beck and Wade (2004) and Suellentrop (2007).

33. Suellentrop (2007:62–63).

34. See Douglas and Hargadon (2004) and Csikszentmihalyi (1977).

35. Johnson (2005); see also Johnson (2006) and Beck and Wade (2004).

36. See Danet (1997) and Csikszentmihalyi (1977). See also Turkle (1984) on aspects of "flow" created as teenagers engage in computer games.

37. Douglas and Hargadon (2004) and Juul (2005:112–15).

38. Aarseth (2006:42).

39. Fallows (2006), Kendall (2002), and Steinkuehler and Williams (2006).

40. See Ling (2004a), Castells et al. (2004), Lenhart et al. (2005 and 2001), and Lenhart and Madden (2006). Of course, we also use these applications for other reasons, especially self-expression (see chapter 8).

41. Oldenburg (1999); see Steinkuehler and Williams (2006).

42. Steinkuehler and Williams (2006).

43. Oldenburg (1999:32).

44. See Ling (2005a and 2004a) and Castells et al. (2004).

45. In Rainie (2006).

46. See Nonnecke and Preece (2000), Preece (2000), Ridings and Gefen (2004).

47. As noted and discussed in Preece (2000:89); see also Nonnecke and Preece (2000 and 2001). On the high proportion of community members that may be lurkers, see Nonnecke and Preece (2000 and 2001), Baym (2000), and Preece (2000:87–90).

48. See Ridings and Gefen (2004), Baym (2000), and Nonnecke and Preece (2000 and 2001).

49. See Steinkuehler and Williams (2006).

50. See Baym (1995b) and Ferrara, Brunner, and Whittemore (1991).

51. Simmel (1964:9–10).

52. See Brown and Levinson ([1978] 1987), Bretag (2006), Baker (2005:35–36), and Baym (1995b). On the creation of common ground, see Clark and Brennan (1993) and Preece (2000: 156–66).

53. Burnett and Buerkle (2004).

54. See Baym (1995b), Bretag (2006), Norrick (1993), and Palmer (1994).

55. Mulkay (1988:69); see also Baym (1995b).

56. Baym (1995b).

57. Fox (2001). See also Geser (2004), Palen, Salzman, and Youngs (2000), and Licoppe and Heurtin (2002), and chapter 8 herein, on social "grooming" via portable technology.

58. See Ben-Ze'ev (2004:145–46).

59. As noted in her research by Fox (2001), though sociologist Lori Kendall found considerable sharing of "masculine intimacies" between men in the online "virtual pub" MUD she studied (2002).

60. Fox (2001), see also Ling (2004a) and Castells et al. (2004).

61. Fox (2001).

62. See Ben-Ze'ev (2004), Baker (2005), and Whitty (2005).

63. See Whitty in Baker (2005:xi–xii) and Whitty (2005).

64. See Ben-Ze'ev (2004:149–50), Chayko (2006), and, again, Whitty in Baker (2005:xi).

65. As noted in Hardey (2002:578).

66. See Thurlow et al. (2004:578) and Baker (2005).

67. On the nature of secrets and relationships, see Nippert-Eng (2006). See also Petronio (2002) and Simmel (1906).

68. In Khazan (2006).

69. Ibid.

70. On internet dependency and "addiction" to a variety of behaviors online see LaRose et al. (2001), Aviram and Amichai-Hamburger (2005), Burnett and Buerkle (2004), Suhail and Bargees (2006), Khazen (2006), and Lomrantz (2006). On the compulsive buying of consumer goods online, see Dittmar, Long, and Bond (2007). On online gambling, see Wood and Williams (2007). On overuse of internet pornography in particular see Hanus (2006), Paul (2005), and Cassel and Pringle (2005) on the increasing use of mobile phones for sexualized content.

71. Hanus (2006:60).

72. See Morahan-Martin (2005), Sanders et al. (2000:241), and LaRose et al. (2001).

73. In Khazan (2006). Anderson and Rainie (2006) present contrasting views on this.

74. See LaRose et al. (2001).

75. In Khazan (2006).

76. See Suhail and Bargees (2005) and LaRose et al. (2001).

77. See Spitzberg (2006).

78. Griffiths (2001:3).

79. Noted in Flora (2007). See also Turkle in Else (2007).

80. Yee's study is described in Khazan (2006). For Wood and Williams's study, see Wood and Williams (2007).

81. See Wellman et al. (2006:24), Wellman and Haythornthwaite (2002), and Chayko (2002).

82. See Thornburgh (2006) on the parental tolerance of gambling as a social activity for children, and Castells et al. (2004:159) on consumerism.

83. Hanus (2006:60), see also Paul (2005).

84. Hanus (2006:60).

CHAPTER 5. SOCIAL NETWORKING: CONVENIENCE, PRACTICALITY, AND SOCIABILITY

1. See Chayko (2002, especially pp. 108–12).

2. Wellman and Hampton (1999).

3. See boyd (2006a and 2006b).

4. Fortunati (2000). On the "always-on mode," see Quan-Haase and Wellman (2002), Quan-Haase, Cothrel, and Wellman (2005), and Gray et al. (2003). For more on grooming talk, see Fox (2001), Wei and Lo (2006), Ling (2004a), Ito and Okabe (2005:264–66), Nakajima, Himeno, and Yashi (1999), Flora (2007). See Reid and Reid (2004:115) on "text circles" of friendship groups who almost perpetually text message one another, and Miyata et al. (2005) and Geser (2004) on "friendship circles" connected by mobile phone. Chapters 4 and 6 of the current work discuss grooming talk and perpetually open network pathways in additional detail.

5. See Dodson (2006).

6. Wellman and Hampton (1999), see also Hampton (2003:106), Kavanaugh et al. (2005), Kavanaugh and Patterson (2002), and Kayahara and Wellman (2007).

7. Post to Association for Internet Researchers mailing list, May 3, 2005, expanded upon in personal email correspondence with me on September 29, 2006. See also Geser (2004) and Fox (2001).

8. See, for example, Boase et al. (2006), Chayko (2007), Kumar et al. (2005:112), and Ishii (2006:109) on how text messaging in Japan tends to reinforce existing friendships rather than create new ones.

9. See Bryant, Sanders-Jackson, and Smallwood (2006:114), Lenhart, Madden, and Hitlin (2005), and Lenhart, Rainie, and Lewis (2001).

10. See Zhao (2006).

11. Rheingold (2002:195), providing a description and overview of Wellman's views on "networked individualism." See also Geser (2004) and all the Wellman and Wellman et al. sources.

12. Wellman (2002).

13. See, for example, Weinberg et al. (1996), Brennan, Moore, and Smyth (1992), Abell and Galinsky (2002), Griffiths (2001), Bryant et al. (2006), Ling (2004a and 2004b), Castells et al. (2004), and Geser (2004).

14. See Hargittai (2002).

15. See Sakkopoulos, Lytras, and Tsakalidis (2006:135).

16. See, for example, Katz and Aackhus (2002), Ling (2004a, 2004b, 2005a, and 2005b), Castells et al. (2004), and Geser (2004).

17. See Galston (1999) and Williams (2006).

18. Ling (2004a:74); see also Geser (2004) and Fortunati (2000).

19. Geser (2004:7).

20. Ling (2004a:69–79). See also Ling and Yttri (2002), and for additional examples of this, Eldridge and Grinter (2001) and Castells et al. (2004).

21. See Rheingold (2002:193–94).

22. On the "softening" of time and schedules, see Ling (2004a:73–75), Ling and Yttri (2002), Rheingold (2002:193–95), and Geser (2004).

23. On email copying, see Skovholt and Svennevig (2006). For more on multitasking, see Wellman et al. (2006), Grinter et al. (2001), Lenhart et al. (2001 and 2005), Nastri et al. (2006), and Quan-Haase et al. (2005).

24. Wellman et al. (2006).

25. Ling (2005a).

26. Madden and Lenhart (2006).

27. See Cooper, McLoughlin, and Campbell (2000), Thurlow et al. (2004:137), Baker (2005), and Whitty (2005).

28. On forms of sexual activity that take place via technology, see Baker (2005), Cooper et al. (2000), Ben-Ze'ev (2004), and Thurlow et al. (2004:141–42), and on antecedents of casual online dating, see Peter and Valkenburg (2007).

29. See Baker (2005:58).

30. See Madden and Lenhart (2006).

31. ScienceDaily.com. On the success of online dating, see McKenna et al. (2002), Griffiths (2001), Galal (2003), Baker (2005), Gibbs, Ellison, and Heino (2006), Lawson and Leck (2006), and ScienceDaily.com.

32. Hollander (2004).

33. Noted in Griffiths (2001:6). On dating without real-time demands, see Galal (2003) and McKenna et al. (2002:19).

34. See Gibbs et al. (2006) and Lawson and Leck (2006).

35. See Baker (2005), ScienceDaily.com, and Thurlow et al. (2004:142).

36. McKenna et al. (2002) as discussed in Galal (2003) and Baker (2005).

37. Baker (2005:151).

38. See Whitty (2005:123), Aviram and Amichai-Hamburger (2005), Young (1999), Young et al. (2000), and Ben-Ze'ev (2004).

39. Aviram and Amichai-Hamburger (2005); see Wiggins and Lederer (1984), Cooper et al. (2000), and Young et al. (2000).

40. See Whitty (2005).

41. Chayko (2007).

42. See Wenger (1998) and Cuthbert, Clark, and Linn (2002).

43. Van Dijk (2005:143).

44. See Rainie (2005).

45. See Burnett and Buerkle (2004).

46. Alexander (2004:31).

47. See Ellis et al. (2006:127). On the use of blogs in teaching, see Du and Wagner (2007).

48. See Young et al. (2000:127), Renninger and Shumar (2002); Haythornthwaite (2002), and Cuthbert et al. (2002).

49. See van 't Hooft and Kelly (2004:131) and Bennett and Fessenden (2006:133).

50. See Guldberg and Pilkington (2006), Cramer et al. (2006), and MacKinnon and Williams (2006).

51. See Renninger and Shumar (2002), Cuthbert et al. (2002), Kling and Courtright (2003).

52. Guldberg and Pilkington (2006:129).

53. See Madden (2006), Rainie (2006), and Wellman et al. (1996).

54. See Skovholt and Svennevig (2006), Schmidt (1997), Schopler et al. (1998), and Sassen (2004).

55. See Rainie (2006). On the impact of gaming on work, see chapter 4.

56. Silver, excerpt from October 2, 2006 blog post at silverinsf.blogspot.com.

57. See Fountain (2005).

58. On the strategies consumers use to determine whether they will risk disclosing personal information during online commercial transactions, see Metzger (2007). For more on the dynamics underlying and influencing such transactions, see Agres, Edberg, and Igbaria (1998), Porter (2004), Madden (2006), and Koyuncu and Bhattacharya (2004).

59. As demonstrated in Kauffmann and Wood (2006).

60. Merton provides this classic theoretical formulation (1957).

61. See Wellman and Gulia (1999), Wellman (1997, 2001), Wellman and Hampton (1999), Wellman et al. (1996), and Haythornthwaite (2002 and 2005).

62. See Chayko (2002), Preece (2000:174), and Caughey (1984).

63. See Granovetter (1973).

64. See Haythornthwaite on connectivity in online social networks (2005) and her earlier work on networks in learning communities (2002).

65. As noted in Haythornthwaite (2005).

CHAPTER 6. BEING THERE: CONSTANT AVAILABILITY

1. See Katz and Aackhus (2002). For various perspectives on the concept of "being there," see Riva and Divide (2003) and, especially, Zhao (2003).

2. See Roloff and Solomon (1989:292).

3. As noted by Kate Fox in Flora (2007:52).

4. As described by Quan-Haase et al. (2005:142–43). See also Quan-Haase and Wellman (2002), Gray et al. (2003), Ito and Okabe (2005), and Nastri, Pena, and Hancock (2006:119).

5. See Ito and Okabe (2005), Fox (2001), and Stone and Richtel (2007). The Graf quote is from Stone and Richtel (2007).

6. Weinberger, excerpt from May 4, 2007 blog post at www.hyperorg. com/backissues/joho-may04-07.html#more

7. Stone, excerpt from December 21, 2006 blog post on WikiHome at continuouspartialattention.jot.home. See also Rainie (2006).

8. See Haddon (2000).

9. Israel (2006).

10. As noted in Reid and Reid (2004). See also Geser (2004) and Flora (2007).

11. Ito and Okabe (2005:262).

12. See also Geser (2004) and Okabe (2004).

13. See Griffiths (2001) and Galal (2003).

14. Berger and Kellner (1964:7; emphasis added).

15. As noted by Flanagin and Metzger (2001:172); see Chayko (2002:142).

16. See Geser (2004), Ling (2004a), Katz and Aackhus (2002), and Castells et al. (2004).

17. See Geser (2004), Palen et al. (2000), Rakow and Navarro (1993), and Else (2007).

18. Geser (2004).

19. As argued in Chayko (2002).

20. As discussed in Ling (2004a), Castells et al. (2004), and Legge (2005).

21. Fox (2001).

22. As noted in Castells et al. (2004) and Palen et al. (2000).

23. Castells et al. (2004:66). See also Ling (2004a:35), Oksman and Rautiainen (2002), and Katz and Aackhus (2002).

24. See Ling (2004a:38–41).

25. Ibid., pp. 46–48. On the mobilization of first responders to emergency situations, see Palen, Hiltz, and Liu (2007).

26. Ibid.

27. Ibid., p. 45.

28. Ibid. On use of the mobile phone as a shield, see also Geser (2004).

29. On this kind of immediacy in communication, see Castells et al. (2004), Agar (2003), Ling (2004a), and Katz and Aackhus (2002). On its consequence for depth and complexity of feeling, see Turkle in Else (2007:11).

30. See Castells et al. (2004:67–69) and Chayko (2002).

31. Legge (2005).

32. Turkle, discussing the "tethering" of the self in Else (2007). See also Hulme and Peters (2001), Castells et al. (2004), and boyd (2006a).

33. On email overload and the role of email in our lives, see Pisello (2007), Whittaker and Sidner (1996), Venolia et al. (2001), Whittaker et al. (2002), CNN. com (2005), and Fisher et al. (2006). Email statistics are from Pisello (2007).

34. See Licoppe and Huertin (2002), Bautsch et al. (2001), and Geser (2004). On gender differences in mobile phone use and the implications of these, see Geser (2004), Puro (2002:23), and Fortunati (2002:51).

35. See Solove (2004).

36. Castells et al. (2004:97).

37. See Ling (2005a, 2005b, and 2004b).

38. See Rosen (2000), Solove (2004), and O'Harrow (2005).

39. Castells et al. (2004:98).

40. See Magid (2007).

41. Rakow and Navarro (1993). See Tugend (2006).

42. See Magid (2007).

43. Castells et al. (2004:161) and Tugend (2006).

44. See Williams and Williams (2005), Tugend (2006), and Castells et al. (2004:175–76).

45. Turkle in Else (2007:11); see Turkle's comments regarding children and portable technologies in Else (2007:10–11).

46. See Castells et al. (2004) and Geser (2004).

47. See Solove (2004) and O'Harrow (2005). On surveillance, the public availability of private information, and technology, see also Lyon (2004), Jenkins (2001), Cassel and Pringle (2005), Wagner (2005), Krueger (2005), and Nordland and Bartholet (2001).

48. In Armour (2006).

49. See McCullagh and Mills (2006) and Page (2006).

50. See McCullagh and Broache (2006).

51. Noted in Armour (2006).

52. Castells et al. (2004:96–98).

53. Ling (2004a:53).

54. See O'Harrow (2005).

55. See Ling (2005a).

56. Fortunati (2000).

57. Israel (2006); Fortunati (2000).

58. See Nippert-Eng (1995), Sheller (2004), Geser (2004), and Rainie (2006).

59. Castells et al. (2004:70–71).

60. Stone, excerpt from December 21, 2006 blog post on WikiHome at continuouspartialattention.jot.home. See also Rainie (2006).

61. Turkle in Else (2007:11).
62. See Licoppe and Huertin (2002) and Bautsch et al. (2001).
63. Geser (2004).
64. Geser (2004).
65. Sheller (2004:39).
66. See Rosen (2004) and Lenhart and Madden (2006).
67. Rosen (2004).
68. Post to Association for Internet Researchers mailing list, September 5, 2005.
69. Hargittai, December 20, 2004 post to www.esztersblog.com.
70. See Ling (2004a).
71. See Raab and Mason (2004)
72. See Raab and Mason (2004), Metzger (2007), and McCaughey and Ayers (2003).
73. Giddens (1984:375).
74. See Chayko (2002).

CHAPTER 7. HARNESSING SOCIAL INTERACTION: THE CONTROL OF TIME, SPACE, AND PEOPLE

1. The ideas that form the basis of this chapter were first developed in Chayko (2005 and 2007). On the ways that society constrains social interaction, see Durkheim ([1893] 1984). On the construction of society "from the ground up," see Berger and Luckmann (1967), Giddens (1984), and Goffman (1983).
2. See CNN.com (2005).
3. On social synchronicity and quasi-synchronicity, see Chayko (2002: 67–73).
4. See, for example, Peter and Valkenburg (2006a) on how adolescents who are socially anxious value highly the controllability of internet communication.
5. See Wellman and Hampton (1999) and Geser (2004).
6. Amichai-Hamburger and McKenna (2006).
7. See Geser (2004) and Bautsch (2001).
8. See Reid and Reid (2004), Baker (2005), and Flora (2007).
9. See Madden (2006), Baker (2005), boyd (2006a), and Lenhart et al. (2005).
10. For more on blogging, see Lenhart and Fox (2006).
11. Blood, personal email correspondence, October 25, 2006.
12. See Baker (2005:68–69).
13. For more on multitasking, see Nastri, Pena, and Hancock (2006), Eldridge and Grinter (2001:37), and Rainie (2006).
14. See Kendall (2002:64–65).
15. For more on portfolios of social connections, see Chayko (2002, especially pp. 6 and 156).
16. See Tugend (2007) and Hallowell (2006).
17. See Quan-Haase et al. (2005), Cross and Parker (2004), and Wellman (1997).

18. Fortunati (2002:55), noted in Geser (2004).

19. See Geser (2004) and chapter 5 of the current work for more on microcoordination and the convenience of portable technologies.

20. On spontaneity in asynchronous communication, see An and Frick (2006). On spontaneity in instant messaging, see Hu et al. (2004). For an excellent overview of humor online and the ways in which it can engender a more spontaneous attitude, see Baym (1995b).

21. See Wellman and Hampton (1999).

22. Kornet (1997).

23. As argued in Nippert-Eng (2006).

24. See Kornet (1997).

25. Fox (2001).

26. See Durkheim ([1893] 1984).

27. Fox (2001).

CHAPTER 8. CREATING, EXPRESSING, AND EXTENDING THE SELF (AND WATCHING OTHERS DO SO)

1. Goffman (1959); see Ellison et al. (2006).

2. In Calvert (2002); see also Huffaker and Calvert (2005).

3. Gee (2001); see also Thomas (2006).

4. See, for example, Mead (1934) and Cooley ([1922] 1964).

5. Donath (1998:29).

6. On processes of socialization, see again, Mead (1934) and Cooley ([1922] 1964).

7. This classic theory of "taking the role of the others" is developed in Mead (1934).

8. Also classic, this theory of the "looking glass self" is developed by Cooley ([1922] 1964).

9. See Shibutani (1955).

10. For examples of this, see Thomas (2006), Huffaker and Calvert (2005), boyd (2006a), and Livingstone et al. (2005).

11. As argued in Hulme and Peters (2001).

12. In Castells et al. (2004:160–61).

13. See Katz and Sugiyama (2006), Castells et al. (2004), and Ling (2004a).

14. See Katz and Sugiyama (2006).

15. Lenhart and Madden (2006); see also Huffaker and Calvert (2005).

16. See Brignall and Van Valey (2005), McMillan and Morrison (2006), Livingstone, Bober, and Helsper (2005), Castells et al. (2004), and Ling (2004a).

17. See Graya et al. (2005).

18. See Bennett and Fessender (2006) and Livingstone et al. (2005:161).

19. See Valkenberg, Schouten, and Peter (2005), Matsuba (2006:166), Thomas (2006), and Castells et al. (2004).

20. See Peter and Valkenburg (2006b).

21. Thomas (2006).

22. boyd (2006a). See boyd (2006a and 2006b) on the meaning of social networking sites and interactions on these sites for youth.

23. Ibid.

24. In Brockman (1996).

25. See boyd (2006a and 2006b), Ling (2004a), Castells et al. (2004), and Else (2007).

26. See Lenhart et al. (2005 and 2001), Bryant et al. (2006), boyd (2006a), and Israel (2006).

27. As noted in Eldridge and Grinter (2001:37), Miyata et al. (2005), and Rainie (2006).

28. Quote from Ling (2004a:111). On teens demonstrating their popularity in technology use, see Ling (2004a:103). On technology enhancing the interaction of groups of teens and adolescents, see Miyata et al. (2005:146), Reid and Reid (2004), and Ling (2004a).

29. See boyd (2006a).

30. Huffaker and Calvert (2005). On blogging and internet content creation among teens and adolescents, see also Lenhart et al. (2005) and Lenhart and Madden (2006).

31. boyd (2006a).

32. See Lenhart et al. (2005), Katz and Sugiyama (2006), Hulme and Peters (2001), Legge (2005), Castells et al. (2004), and Ling (2004a).

33. See Ling (2004a) and Baker (2005).

34. See Psychology Today (2007:14).

35. Legge (2005).

36. See Abell and Galinsky (2002), Brockman (1996), Rainie (2006), Lenhart and Madden (2006), and Lenhart et al. (2005).

37. See Castells et al. (2004:162) and Tugend (2006).

38. See McMillan and Morrison (2006), Lomrantz (2006), Tugend (2006), and Willard (2007).

39. See Miyata et al. (2005:146).

40. As noted in Mesch and Talmud (2006) and Matsuba (2006).

41. Turkle in Brockman (1996).

42. See Grotevant (1998) and Erikson (1993).

43. See Suler (2004).

44. Galal (2003); see also McKenna et al. (2002).

45. Blood (2000b:13–15). On the popularity of blogging, see Lenhart and Fox (2006).

46. Blood (2000b:xi).

47. Barrett (2002:28–29).

48. See Lenhart and Fox (2006).

49. These quotes are all from Blood (2002b:13–15). See also her analyses in (2002a and 2002b) and on her blog at rebeccablood.net.

50. See Baker (2005).

51. As noted in Reid and Reid (2004).

52. As noted in Castells et al. (2004, see, especially, p. 81).

53. See Lin (1999:181) for a discussion of how queer identity is performed online. See also Huffaker and Calvert (2005).

54. Plant (1997:46).

55. See Turkle (1995). On the altering of race and gender online, see, for example, Nakamura (1999), O'Brien (1999), Burkhlater (1999), and Silver (2000).

56. On the riskiness of identity "bending" online, see Preece (2000:154–55), Ling (2004a:164–65), Baker (2005:82–83), and Kendall (2002). On gender and linguistic, conversational, and interactional styles, see Tannen (1991).

57. See Rogers (1951).

58. See Reid and Reid (2004).

59. As McKenna et al. (2002) found.

60. Bargh et al. (2002:34).

61. See Mehra and Merkel (2004), Mitra (2005), Lin (2006), and Baker (2005).

62. Friedman (2005:1). See also McKenna and Bargh (1998).

63. Schrock et al. (2004:66).

64. These quotes are all from Schrock et al. (2004:66).

65. Bowker and Tuffin (2002:329).

66. See Goggin and Newell (2003), Williamson et al. (2001), and Thoreau (2006).

67. As noted in a thorough overview of this issue in Thoreau (2006).

68. See Guo, Bricout, and Huang (2005).

69. As noted in Williamson et al. (2001).

70. As discussed in Geser (2004).

71. See Morahan-Martin and Schumacher (2003:665), Spitzberg (2006), and Matsuba (2006).

72. Wei and Lo (2006). See also Geser (2004), Baker (2005), Reid and Reid (2004), McKenna et al. (2002), and Preece (2000).

73. Wellman and Hampton (1999).

74. For examples of this, see Dean (1999).

75. See Glaser, Dixit, and Green (2002:22) and Carmichael (2003).

76. Lee (2006).

77. Kelly (2005).

78. See Nonnecke and Preece (2001).

79. In Cassidy (2006:54).

80. Barrett (2002:30); see also Lenhart and Fox (2006).

81. Gregg (2006:147).

82. Blood (2002b:10–11).

83. For a discussion of the negotiation of these audiences and self-presentation with regard to them, see boyd and Heer (2006).

84. See Kendall (2002:126) and Geser (2004).

85. boyd and Heer (2006).

86. See Walther and Parks (2002).

87. See Donath (1999), Ramirez et al. (2002), and Ellison et al. (2006).

88. See Donath (1999).

89. See Kendall (2002) and Baker (2005).

90. See Kibby (1997) and Huffaker and Calvert (2005).

91. See Baker (2005) and Walther et al. (2005).

92. See Lenhart et al. (2005), Lenhart and Madden (2006), and Nastri et al. (2006).

93. See Katz and Sugiyama (2006), Castells et al. (2004), and Ling (2004a).

94. See Walther and Tidwell (1995), Ling (2004a), Baker (2005), and Ellison et al. (2006).

95. See Preece (2000), Gibbs et al. (2006), and Ellison et al. (2006).

96. As noted by McKenna in Galal (2003).

97. See Hardey (2002:579), Galal (2003), Gibbs et al. (2006), Ellison et al. (2006), and Baker (2005).

98. Blood, personal email correspondence, October 25, 2006.

99. Goffman (1971:384).

100. See Geser (2004), Gillard (1996).

101. See Gee (2001:99), Stone (1996), and Turkle (1995).

102. Turkle in Brockman (1996).

103. For examples of this, see O'Brien (1999), Burkhlater (1999), and Schrock et al. (2004).

104. See Kendall (2002:222–23).

105. As Huffaker and Calvert (2005) found.

106. Except in the case of multiple personality disorder. On the singularity of identity, see Erikson (1993) and Kendall (2002:224).

107. See Zerubavel (1991) on mental flexibility and fluidity; see also Chayko (2002).

108. Simmel ([1908] 1962). On how we manage these roles and networks, see also Geser (2004), Rheingold (2002), Chayko (2002) and all the Haythornthwaite and Wellman pieces, especially Wellman (2001 and 2002).

109. See van Dijk on the operational skills divide (2005:79–81). See also Rainie (2006), Lenhart and Madden (2006), Lenhart et al. (2005), Suellentrop (2006), Beck and Wade (2004), and Thomas (2006).

110. In Rainie (2006).

111. Thomas (2006). Floridi (2007) also argues that the threshold between the online and offline will soon disappear.

112. Wilson and Atkinson (2005).

113. In Weslander (2006).

114. On "the flexible mind" and flexible, nonbinary forms of social categorization, see Zerubavel (1991) and Chayko (2002). See also Huffaker and Calvert (2005).

CHAPTER 9. SHAPING THE SOCIAL LANDSCAPE: EQUALITIES, INEQUALITIES, POSSIBILITIES

1. See Thurlow et al. (2004:84) and van Dijk (2005).

2. See Guillén and Suárez (2005).

3. See Thurlow et al. (2004:87) and Jordan (2001).

4. Jordan (2001:1); see also Jordan (1999).

5. As noted in Sproull and Kiesler (1993:108–11); see McLaughlin et al. (1995) and Jordan (2001).

6. Dean (1999:1072).

7. As noted in Thurlow et al. (2004:85) and Agres, Edberg, and Igbaria (1998). See also Howard (2007).

8. See Ling (2004a), Levinson (2004:141), and Robbins and Turner (2002).

9. See Danowitz, Wassef, and Goodman (1995), Agres et al. (1998), and Vroman (2007).

10. See Thurlow et al. (2004:87), Dean (1999), and Hargittai and Shafer (2006). On diffusion of the internet by income in the United States as compared to Europe, see Martin and Robinson (2007).

11. See Thurlow et al. (2004:87), Jordan (2001 and 1999), Losh (2004), and McQuaid, Lindsay, and Greig (2004).

12. See Pitts (2004), Thoreau (2006), and Madge and O'Connor (2006).

13. See Jordan (2001 and 1999).

14. See Ferlander and Timms (2006) and Cho et al. (2003).

15. As Kang, Bagchi-sen, and Rao (2005) demonstrate.

16. See Willis and Tranter (2006) and van Dijk (2005).

17. As noted in van Dijk (2005:142) and Thurlow et al. (2004:87).

18. See Jackson et al. (2004), Hoffman and Novak (1998), and McQuaid et al. (2004).

19. See van Dijk (2005), Howard (2007), and Katz and Rice (2002).

20. See Willard (2007).

21. See Jewkes and Andrews (2005).

22. See Castronova (2006).

23. See Biegel (2003).

24. For more on this, see Barak (2005).

25. See Legge (2005), Katz and Rice (2002), and Ling (2004a).

26. See Magid (2001), Willard (2007), National Childrens' Home (2005), Legge (2005), and Ling (2004a).

27. boyd (2006a).

28. See Donath (1998:373).

29. Glaser et al. (2002).

30. For examples, see Monk et al. (2004), Geser (2004), Ling (2004a), Campbell (2006), and Levinson (2004).

31. Examples of this can be found in Campbell (2006), Miyata et al. (2005), Miyaki (2005), Lenhart et al. (2005 and 2001), Ling (2004a), and Rainie (2006).

32. See Baker (2001) and Thompson (1998:98).

33. boyd (2006a).

34. Noted in Hine (2000:104–105); see boyd (2006a).

35. boyd (2006a).

36. Noted in Baker (2001), see Hall et al. (1978), and Goode and Ben-Yehuda (1994).

37. Dean (1999:1093).

38. See Herring et al. (2002) and Dean (1999). On online regulation and policing, see Williams (2007).

39. In Hewitt (2006:121).

40. See Chayko (2002:133–35) and Coughlin (2001).

41. See McCaughey and Ayers (2003:6–14).

42. In Kidd (2003).

43. See Lemann (2006) and McCaughey and Ayers (2003).

44. As cited in Sterling (1992:86), noted in Jordan (2001).

45. See Pickard (2006).

46. See Kavanaugh et al. (2005) and Kavanaugh and Patterson (2002); see also Haythornthwaite (2005), and Vroman (2007).

47. See Krueger (2005).

48. See Goodwin et al. (2001:6).

49. See Schrock et al. (2004) for more on emotion and consciousness-raising, and for more examples like these of emotionality in social movements.

50. See Bacon (2006), Shah et al. (2005), and Lemann (2006).

51. Yu (2006:303).

52. DemocracyNow.org (2006).

53. Kahn and Kellner (2004).

54. Noted by David Silver, post to Association for Internet Researchers mailing list, April 12, 2006.

55. As noted, along with other such examples, by Bacon (2006).

56. See Kulikova and Perlmutter (2007).

57. See Kahn and Kellner (2004:87–88).

58. Langman (2005:42).

59. Hands (2006).

60. McCaughey and Ayers (2003:2).

61. Bacon (2006); see also Kavada (2005).

62. Silver, post to Association for Internet Researchers mailing list, April 12, 2006.

63. Silver (2000) makes this argument most persuasively in his excellent overview of the study of cyberculture and the importance of "critical cyberculture studies." See also Silver (2003) and Silver and Massaneri (2006).

64. Kelly (2005).

65. Grossman (2006).

66. See Grossman (2006).

67. As noted in Bargh (2002).

68. In Rainie (2006).

69. As argued in Putnam (1995).

70. Wellman and Hampton (1999).

71. As noted in Chayko (2007). See also Boase et al. (2006), Jackson et al. (2004), Kavanaugh and Patterson (2002), Hampton and Wellman (2002), Wagner et al. (2002), Rheingold (2000 and 2002).

72. See Calhoun (1986), Chayko (2002), Walther (1996), Turner et al. (2001).

73. See Chayko (2003).

74. See Anderson and Tracey (2002).
75. See Baym et al. (2004) and Wellman et al. (2006).
76. As W. I. Thomas taught, in Thomas and Thomas (1928).
77. See Wellman and Haythornthwaite (2002) and Chayko (2003).
78. Wellman et al. (2006).

APPENDIX I. THE METHODOLOGY

1. See Persichitte, Young, and Tharp (1997) and Chayko (2002). Much of the methodology for the current study, and the description of the methodology here, has been adapted from Chayko (2002).
2. Persichitte et al. (1997).
3. See Meho (2006).
4. See Mann and Stewart (2000:128).
5. See Meho (2006).
6. See Persichitte et al. (1997).
7. Persichitte et al. (1997); see also Hewson et al. (2003:51).
8. Persichitte et al. (1997).
9. See Hewson et al. (2003:45–46).
10. See Baym (2006:83).
11. See Mann and Stewart (2000:128).
12. See Mann and Stewart (2000:132–44).
13. See Chayko (2002) and Mann and Stewart (2000:130–31).
14. See Baym (1995) for an overview of this literature and well-framed guidelines regarding computer-mediated research, and Mann and Stewart (2000) for a more recent discussion.

REFERENCES

Aarseth, E. 2006. "How We Became Postdigital: From Cyberstudies to Game Studies." In Silver, D., and Massanari, A. (eds.), *Critical Cyberculture Studies*. New York: New York University Press.

Abell, M. L., and Galinsky, M. J. 2002. "Introducing Students to Computer-Based Group Work Practice." *Journal of Social Work Education*. 38:1:39–54.

Adams, C., and Avison, D. E. 2003. "Dangers Inherent in the Use of Techniques: Identifying Framing Influences." *Information, Technology & People*. 16:2:203–34.

Agres, C., Edberg, D., and Igbaria, M. 1998. "Transformation to Virtual Societies: Forces and Issues." *Information Society*. 14:2:71–83.

Alavi, M., and Leidner, D. 2001. "Knowledge Management and Knowledge Management Systems: Conceptual Foundations and Research Issues." *MIS Quarterly*. 25:1:107–36.

Alexander, B. 2004. "Going Nomadic: Mobile Learning in Higher Education." *EDUCAUSE Review*. 39:5:28–35.

Amichai-Hamburger, Y., and McKenna, K. Y. A. 2006. "The Contact Hypothesis Reconsidered: Interacting via the Internet." *Journal of Computer-Mediated Communication*. 11:3:7. Retrieved November 20, 2006, at http://jcmc.indiana.edu/vol11/issue3/amichai-hamburger.html

An, Y. J., and Frick, T. 2006. "Student Perceptions of Asynchronous Computer-Mediated Communication in Face-to-Face Courses." *Journal of Computer Mediated Communication*. 11:2:5. Retrieved November 20, 2006, at http://jcmc.indiana.edu/vol11/issue2/an.html

Anderson, B. 1983. *Imagined Communities*. London: Thetford Press.

Anderson, B. 1991. *Imagined Communities: Reflections on the Origin and Spread of Nationalism*. New York, NY: Verso.

Anderson, B., and Tracey, K. 2002. "Digital Living: The Impact (or Otherwise) of the Internet on Everyday British Life." In Wellman, B. and Haythornthwaite, C. (eds.) *The Internet in Everyday Life*. Malden, MA: Blackwell Publishing.

Anderson, J., and Rainie, L. 2006. "The Future of the Internet II." *Pew Internet and American Life Project*. Retrieved September 24, 2006, at www.pewinternet.org

Armour, S. 2006. "Keeping an Eye on You." *Home News Tribune.* November 27: D1.

Atkinson, M. 2004. "Tattooing and Civilizing Processes: Body Modification as Self-Control." *Canadian Review of Sociology & Anthropology.* 41:2:125–46.

Aviram, I., and Amichai-Hamburger, Y. 2005. "Online Infidelity: Aspects of Dyadic Satisfaction, Self-Disclosure, and Narcissism." *Journal of Computer-Mediated Communication.* 10:3:1. Retrieved November 20, 2006, at http://jcmc.indiana.edu/vol10/issue3/aviram.html

Bacon, P. 2006. "The Netroots Hit Their Limit." *Time.* September 24. Retrieved September 30, 2006, at http://www.time.com/time/magazine/article/0,9171,1538663,00.html

Baker, A. 2005. *Double Click: Romance and Commitment Among Online Couples.* Cresskill, NJ: Hampton Press.

Baker, P. 2001. "Moral Panic and Alternative Identity Construction in Usenet." *Journal of Computer-Mediated Communication.* 7:1 Retrieved September 24, 2007, at http://jcmc.indiana.edu/vol7/issue1/baker.html

Barabási, A. L. 2003. *Linked: How Everything Is Connected to Everything Else and What It Means.* New York: Plume.

Barak, A. 2005. "Sexual Harassment on the Internet." *Social Science Computer Review.* 23:1:77–92.

Barak, A., and Wander-Schwartz, M. 2000. "Empirical Evaluation of Brief Group Therapy Conducted in an Internet Chat Room." *Journal of Virtual Environments.* Retrieved May 23, 2005, at http://www.brandeis.edu/pubs/jove/HTML/V5/CHERAPY3.HTM

Barbatsis, G., Fegan, M., and Hansen, K. 1999. "The Performance of Cyberspace: An Exploration into Computer-Mediated Reality." *Journal of Computer-Mediated Communication.* Retrieved June 28, 2006, at http://www.ascusc.org/jcmc/volS/issue1/barbatsis.html

Bargh, J. A. 2002. "Beyond Simple Truths: The Human-Internet Interaction." *Journal of Social Issues.* 58:1:1–8.

Bargh, J. A., McKenna, K. Y. A., and Fitzsimmons, G. M. 2002. "Can You See the Real Me? Activation and Expression of the 'True Self' on the Internet." *Journal of Social Issues.* 58:1:33–49.

Barlow, J. P., Birkets, S., Kelly, K., and Slouka, M. 1995. "What Are We Doing On Line." *Harper's.* August: 35–46.

Barrett, C. 2002. "Anatomy of a Weblog." In Perseus (eds.), *We've Got Blog.* Cambridge, MA: Perseus.

Bateson, G. 1972. "A Theory of Play and Fantasy." In Bateson, G. (ed.), *Steps to an Ecology of Mind.* New York: Ballantine.

Baumann, Z. 2003. *Liquid Love: On the Frailty of Human Bonds.* Cambridge, UK: Polity.

Bautsch, H., Granger, J., Karnjate, T., Kahn, F., Leveston, Z., Niehus, G., et al. 2001. "An Investigation of Mobile Phone Use: A Socio-Technical Approach." Department of Industrial Engineering, University of Wisconsin—Madison. Retrieved October 16, 2006, at http://www.cae.wisc.edu/~granger/IE449/IE449_0108.pdf

Baym, N. 1995a. "The Emergence of Community in Computer-Mediated Communication." In Jones, S. (ed.), *Cybersociety: Computer-Mediated Communication and Community*. Thousand Oaks, CA: Sage.

Baym, N. 1995b. "The Performance of Humor in Computer-Mediated Communication." *Journal of Computer-Mediated Communication*. 1:2. Retrieved October 2, 2006, at http://jcmc.indiana.edu/vol1/issue2/baym.html

Baym, N. 2000. *Tune In, Log On: Soaps, Fandom and Online Community*. Thousand Oaks, CA: Sage.

Baym, N. 2006. "Finding the Quality in Qualitative Research." In Silver, D., and Massanari, A. (eds.), *Critical Cyberculture Studies*. New York: New York University Press.

Baym, N., Zhang, V. B. and Lin, M. C. 2004. "Social Interactions Across Media." *New Media & Society*. 6:3:299–318.

Beck, J. C. and Wade, M. 2004. *Got Game: How the Gamer Generation Is Reshaping Business Forever*. Boston, MA: Harvard Business School Publishing.

Beder, J. 2005. "Cybersolace: Technology Built on Emotion." *Social Work*. 50:4:355–58.

Bell, C., and Newby, H. 1973. *Community Studies: An Introduction to the Sociology of the Local Community*. New York: Praeger.

Bennett, L., and Fessenden, J. 2006. "Citizenship Through Online Communication." *Social Education*. 70:3:144–46.

Ben-Ze'ev, A. 2004. *Love Online: Emotions on the Internet*. Cambridge: Cambridge University Press.

Berger, P., and Kellner, H. 1964. "Marriage and the Construction of Reality." *Diogenes*. 45:1–25.

Berger, P., and Luckmann, T. 1967. *The Social Construction of Reality*. Garden City, NY: Doubleday.

Biegel, S. 2003. *Beyond Our Control: Confronting the Limits of Our Legal System in the Age of Cyberspace*. Cambridge, MA: MIT Press.

Bird, J. 2003. " 'I Wish To Speak to the Despisers Of The Body': The Internet, Physicality, and Psychoanalysis." *Journal for the Psychoanalysis of Culture and Society*. 8:1:121–27.

Blackler, F. 1995. "Knowledge, Knowledge Work and Organizations: An Overview and Interpretation." *Organization Studies*. 16:6:1021–46.

Blood, R. 2002a. "Introduction." In Perseus (eds.), *We've Got Blog*. Cambridge, MA: Perseus.

Blood, R. 2002b. "Weblogs: A History and Perspective." In Perseus (eds.), *We've Got Blog*. Cambridge, MA: Perseus.

Boase, J., Horrigan, J., Wellman, B., and Rainie, L. 2006. "The Strength of Internet Ties." *Pew Internet and American Life Project*. Retrieved March 15, 2006, at www.pewinternet.org

Bodmer, C. 2001. "Women Are the Majority of Online Gamers." *Marketing to Women*. 14:1:12.

Bodnar, J. 1992. *Remaking America: Public Memory, Commemoration, and Patriotism in the Twentieth Century*. Princeton, NJ: Princeton University Press.

Boland, R. J., and Tenkasi, R. V. 1995. "Perspective Making and Perspective Taking in Communities of Knowing." *Organization Science*. 6:4:350–72.

Boland, R. J., Tenkasi, R. V., and Te'eni, D. 1994. "Designing Information Technology to Support Distributed Cognition." *Organization Science*. 5:3:456–75.

Bourdieu, P. 1983. "The Forms of Capital." In Richardson, J. G. (ed.), *Handbook of Theory and Research for the Sociology of Education*. Westport, CT: Greenwood Press.

Bowker, N., and Tuffin, K. 2002. "Disability Discourses for Online Identities." *Disability and Society*. 17:3:327–44.

boyd, d. 2006a. "Identity Production in a Networked Culture: Why Youth Heart MySpace." Presentation for the American Association for the Advancement of Science, February 19, St. Louis, MO.

boyd, d. 2006b. "Friends, Friendsters, and MySpace Top 8: Writing Community into Being on Social Network Sites." *First Monday*. December: 11:12. Retrieved November 10, 2006, at http://www.firstmonday.org/issues/issue11_12/boyd/index.html

boyd, d., and Heer, J. 2006. "Profiles as Conversation: Networked Identity Performance on Friendster." In Proceedings of the Hawaii International Conference on System Sciences (HICSS-39), Persistent Conversation Track. Kauai, HI: IEEE Computer Society. January 4–7.

Brennan, F., Moore, S. M., and Smyth, K. A. 1992. "Alzheimer's Disease Caregivers' Uses of a Computer Network." *Western Journal of Nursing Research*. 14:662–73.

Bretag, T. 2006. "Developing 'Third Space' Interculturality Using Computer-Mediated Communication." *Journal of Computer-Mediated Communication*. 11:4:5. Retrieved October 2, 2006, at http://jcmc.indiana.edu/vol11/issue4/bretag.html

Brignall, T., and Van Valey, T. 2005. "The Impact of Internet Communications on Social Interaction." *Sociological Spectrum*. 25:3:335–48.

Brockman, J. 1996. *Digerati: Encounters with the Cyber Elite*. Hard Wired Books. Chapter 31. Retrieved November 15, 2006, at http://www.edge.org/documents/digerati/Turkle.html

Broom, A. 2005. "Prostate Cancer, Masculinity, and Online Support as a Challenge to Medical Expertise." *Journal of Sociology*. 41:1:87–104.

Brown, J. S., and Duguid, P. 1994. "Borderline Issues: Social and Material Aspects of Design." *Human-Computer Interaction*. 9:1:3–36.

Brown, J. S., and Duguid, P. 2000. "Knowledge and Organization: A Social-Practice Perspective." *Organization Science*. 12:2:198–213.

Brown, P. and Levinson, S. C. [1978] 1987. *Politeness: Some Universals in Language Usage*. Cambridge: Cambridge University Press.

Browne, S. H. 1995. "Reading, Rhetoric and Texture of Public Memory." *Quarterly Journal of Speech*. 81:2:237–49.

Bryant, J. A., Sanders-Jackson, A., and Smallwood, A. M. K. 2006. "IMing, Text Messaging, and Adolescent Social Networks." *Journal of Computer-Mediated Communication*. 11:2:10. Retrieved October 2, 2006, at http://jcmc.indiana.edu/vol11/issue2/bryant.html

Buckland, M. K. 1991. "Information as Thing." *American Society for Information Science*. 42:5:351–60.

Burgoon, J. K., and Hale, J. L. 1987. "Validation and Measurement of the Fundamental Themes of Relational Communication." *Communication Monographs*. 54:1:9–41.

Burkhlater, B. 1999. "Reading Race Online." In Smith, M. A., and Kollock, P. (eds.), *Communities in Cyberspace*. New York: Routledge.

Burnett, G. 2002. "The Scattered Members of an Invisible Republic: Virtual Communities and Paul Ricoeur's Hermeneutics." *The Library Quarterly Chicago*. 72:2:155–78.

Burnett, G., and Buerkle, H. 2004. "Information Exchange in Virtual Communities: A Comparative Study." *Journal of Computer-Mediated Communication*. 9:2. Retrieved October 16, 2006, at http://jcmc.indiana.edu/vol9/issue2/burnett.html

Burt, R. S. 1997. "The Contingent Value of Social Capital." *Administrative Science Quarterly*. 42:339–65.

Calhoun, C. 1998. "Community Without Propinquity Revisited: Communications Technology and the Transformation of the Urban Public Sphere." *Sociological Inquiry*. 68:3:373–97.

Calvert, S. L. 2002. "Identity Construction on the Internet." In Calvert, S. L., Jordan, A. B., and Cocking, R. R. (eds.), *Children in the Digital Age: Influences of Electronic Media on Development*. Westport, CT: Praeger.

Campbell, S. 2006. "Perceptions of Mobile Phones in College Classrooms: Ringing, Cheating, and Classroom Policies." *Communication Education*. 55:3:280–94.

Campbell, S., and Kelley, M. 2006. "Mobile Phone Use in AA Networks: An Exploratory Study." *Journal of Applied Communication Research*. 34:2:191–208.

Cannon Bowers, J. A., and Salas, E. 2001. "Reflections on Shared Cognition." *Journal of Organizational Behavior*. 22:2:195–202.

Carlile, P. R. 2002. "A Pragmatic View of Knowledge and Boundaries." *Organization Science*. 13:4:442–55.

Carlson, A. C., and Hocking, J. E. 1988. "Strategies of Redemption and the Vietnam Veterans' Memorial." *Western Journal of Speech Communication*. 52:203–15.

Carmichael, P. 2003. "The Internet, Information Architecture and Community Memory." *Journal of Computer-Mediated Communication*. 8:2. Retrieved September 13, 2006, at http://jcmc.indiana.edu/vol8/issue2/carmichael.html

Cascio, T., and Gasker, J. 2001. "Everyone Has a Shining Side: Computer-Mediated Mentoring in Social Work Education." *Journal of Social Work Education*. 37:2:283–94.

Cashel, J. 2001. "Top Ten Trends in Online Communities." Retrieved May 19, 2005, on www.onlinecommunityreport.com

Cassel, B., and Pringle, D. 2005. "Sex Cells: Wireless Operators Find That Racy Cellphone Video Drives Surge in Broadband Use." *The Wall Street Journal*. May 12.

Cassidy, J. 2006. "Me Media." *The New Yorker.* May 15: 50–59.

Castells, M., Fernandez-Ardevol, M., Qiu, J., and Sey, A. 2004. "Social Uses of Wireless Communications: The Mobile Information Society." Paper prepared for the International Workshop on Wireless Communication Policies and Prospects: A Global Perspective, USC, October 8–9.

Castronova, J. R. 2006. "Operation Cyber Chase and Other Agency Efforts to Control Internet Drug Trafficking: The 'Virtual' Enforcement Initiative Is Virtually Useless." *Journal of Legal Medicine.* 27:2:207–24.

Caughey, J. 1984. *Imaginary Social Worlds.* Lincoln: University of Nebraska Press.

Cerulo, K. 1995. *Identity Designs: The Sights and Sounds of a Nation.* New Brunswick, NJ: Rutgers University Press.

Cerulo, K., and Ruane, J. 1996. "Death Comes Alive: Technology and the Re-Conception of Death." *Science as Culture.* 28:1:444–66.

Cerulo, K., and Ruane, J. 1998. "Coming Together: New Taxonomies for the Analysis of Social Relations." *Sociological Inquiry.* 68:3:398–425.

Cerulo, K., Ruane, J., and Chayko, M. 1992. "Technological Ties That Bind: Media-Generated Primary Groups." *Communication Research.* 19:1:109–29.

Charlton, T., Panting, C. and Hannan, A. 2002. "Mobile Telephone Ownership and Usage Among 10- and 11-Year-Olds: Participation and Exclusion." *Emotional Behavioural Difficulties.* 7:3:152–63.

Chayko, M. 1993. "What Is Real in the Age of Virtual Reality? 'Reframing' Frame Analysis for a Technological World." *Symbolic Interaction.* 16:2:171–81.

Chayko, M. 2002. *Connecting: How We Form Social Bonds and Communities in the Internet Age.* Albany, NY: State University of New York Press.

Chayko, M. 2003. "Book Review: The Internet in Everyday Life." *Contemporary Sociology.* 32:6:728–30.

Chayko, M. 2005. "Harnessing Social Interaction: How We Use the Internet to Shape and Control Interpersonal Contact." Paper presented at the annual meeting of the Eastern Sociological Society, Washington, DC, March 20.

Chayko, M. 2006. "Book Review: Love Online." Resource Center for Cyberculture Studies. April. Retrieved May 1, 2006, at http://rccs.usfca.edu/bookinfo.asp?ReviewID=371&BookID=301

Chayko, M. 2007. "The Portable Community: Envisioning and Examining Mobile Social Connectedness." *The International Journal of Web Based Communities.* 3:4:373–85.

Cho, J., De Zuniga, H. G., Rojas, H., and Shah, D.V. 2003. "Beyond Access: The Digital Divide and Internet Uses and Gratifications." *IT and Society.* Stanford University. 4:1:46–72.

Christian, A. 2005. "Contesting the Myth of the 'Wicked Stepmother': Narrative Analysis of an Online Stepfamily Support Group." *Western Journal of Communication.* 69:1:27–47.

Christensen, T. H. 2004. "ICT-Mediated Proximity—Being Far Away and Close Together." Paper presented at the 4S/EASST conference, Paris, France. August.

Clark, H. H., and Brennan, S. E. 1993. "Grounding in Communication." In Baecker, R. M. (ed.), *Readings in Groupware and Computer-Supported Cooperative Work*. San Mateo, CA: Morgan Kaufmann Publishers.

CNN.com. 2005. "Emails Hurt IQ More Than Pot." April 25. Retrieved May 10, 2006, at http://www.cnn.com/2005/WORLD/europe/04/22/text.iq/

Cohen, A. P. 1985. *The Symbolic Construction of Community*. Chichester: Ellis Horwood.

Cook, J., and Brown, J. S. 1999. "Bridging Epistemologies: The Generative Dance Between Organizational Knowledge and Organizational Knowing." *Organization Science*. 10:4:381–400.

Cooley, C. [1922] 1964. *Human Nature and the Social Order*. New York: Schocken.

Cooper, A., McLoughlin, I. P., and Campbell, K. M. 2000. "Sexuality in Cyberspace: Update for the 21st Century." *CyberPsychology and Behavior*. 3:4:521–36.

Coughlin, K. 2001. "Cyber-Predators Lie in Wait for Children." *The Star-Ledger*. June 30: 3.

Cox, J., and Dubie, D. "Experts: Security on Handhelds Far Too Lax." *Network World*. Retrieved March 5, 2006, at http://www.arnnet.com.au/index.php?id=563459033&fp=2&fpid=1

Cox, E., and Leonard, H. 1990. "Weaving Community Links: The Cost Benefits of Telephones in Maintaining the Social Fabric Through the Unpaid Work of Women." *The Distaff Papers, Australia*. Retrieved October 16, 2006, at http://216.92.140.78/hostedpages/Distaff/Telstra/3%20results.htm

Cramer, K., Collins, K., Snider, D., and Fawcett, G. 2006. "Virtual Lecture Hall for In-Class and Online Sections: A Comparison of Utilization, Perceptions, and Benefits." *Journal of Research on Technology in Education*. 38:4:371–81.

Cross, R., and Parker, A. 2004. *The Hidden Power of Social Networks*. Boston, MA: Harvard Business School Press.

Csikszentmihalyi, M. 1977. *Beyond Boredom and Anxiety*. San Francisco: Jossey-Bass.

Curran, K., and Casey, M. 2006. "Expressing Emotion in Electronic Mail." *Kybernetes*. 35:5:616–31.

Curtis, P. 1992. "Mudding: Social Phenomena in Text-Based Social Realities." In Schuler, D. (ed.), *DIAC-92: Directions and Implications of Advanced Computing*. Palo Alto, CA: Computer Professionals for Social Responsibility.

Cuthbert, A. J., Clark, D. B., and Linn, M. C. 2002. "WISE Learning Communities: Design Considerations." In Renninger, K. A., and Shumar, W. (eds.), *Building Virtual Communities: Learning and Change in Cyberspace*. Cambridge: Cambridge University Press.

Dahlberg, L. 2005. "The Corporate Colonization of Online Attention and the Marginalization of Critical Communication." *Journal of Communication Inquiry*. 29:2:160–80.

Dandeker, C. 2004. "The Wars Within: Peoples and States in Conflict." *American Journal of Sociology*. 110:1:232–33.

Danet, B. 1996. "Text as Mask: Gender and Identity on the Internet." Paper presented at the conference on Masquerade and Gendered Identity. Venice.

February 21–24. Retrieved September 20, 2006, at http://atar.mscc.huji. ac.il/~msdanet/mask.html

Danet, B. 1997. "Hmmm . . . Where's That Smoke Coming From? Writing, Play and Performance on Internet Relay Chat." *Journal of Computer-Mediated Communication.* 2:4. Retrieved November 1, 2006, at http://jcmc.indiana. edu/vol2/issue4/danet.html

Danet, B. 2001. *Cyberplay: Communicating Online.* Oxford: Berg.

Danowitz, A. K., Nassef, Y., and Goodman, S. E. 1995. "Cyberspace Across the Sahara: Computing in North Africa." *Communications of the ACM.* 38:12:23–28.

Davis, M. 1983. *Smut.* Chicago: University of Chicago Press.

Dayan, D., and Katz, E. 1992. *Media Events: The Live Broadcasting of History.* Cambridge, MA: Harvard University Press.

Dean, J. 1999. "Virtual Fears." *Journal of Women in Culture and Society.* 24:4:1069–79.

DemocracyNow.org. 2006. "Dallas High School Student Describes Organizing Mass Walkout." April 6. Retrieved October 20, 2006, at http://www. democracynow.org/article.pl?sid=06/04/11/1426244

DiMaggio, P., Hargittai, E., Neuman, W. R., and Robinson, J. P. 2001. "Social Implications of the Internet." *Annual Review of Sociology.* 27:1:307–38.

Dittmar, H., Long, K., and Bond, R. 2007. "When a Better Self Is Only a Click Away: Associations Between Materialistic Values, Emotional and Identity-Related Buying Motives, and Compulsive Buying Tendency Online." *Journal of Social and Clinical Psychology.* 26:4:334–61.

Dodson, S. 2006. "Show and Tell Online." *The Guardian.* March 2. Retrieved September 29, 2006, at http://technology.guardian.co.uk/weekly/ story/0,,1720763,00.html

Donath, J. 1998. "Identity and Deception in the Virtual Community." In Smith, M., and Kollock, P. (eds.), *Communities in Cyberspace.* New York: Routledge.

Donath, J., and boyd, d. 2004. "Public Displays of Connection." *BT Technology Journal.* 22:4:71–82.

Douglas, J. Y., and Hargadon, A. 2004. "The Pleasures of Immersion and Interaction: Schemas, Scripts, and the Fifth Business." In Wardrip-Fruin, N., and Harrigan, P. (eds.), *First Person: New Media as Story, Performance and Game.* Cambridge, MA: MIT Press.

Downs, R. N., and Stea, D. 1977. *Maps in Minds.* New York: Harper.

Drentea, P., and Moren-Cross, J. L. 2005. *Sociology of Health and Illness.* 27:7:920–43.

Du, H. S., and Wagner, C. 2007. "Learning with Weblogs: Enhancing Cognitive and Social Knowledge Construction." *IEEE Transactions on Professional Communication.* 50:1:1–16.

Durkheim, E. [1912] 1965. *The Elementary Forms of Religious Life.* New York: Free Press.

Durkheim, E. [1893] 1984. *The Division of Labor in Sociey.* New York: Free Press.

Eisenhardt, K. M. 1989. "Building Theories from Case Study Research." *Academy of Management Review.* 14:4:532–50.

Eldridge, M., and Grinter, B. 2001. "Studying Text Messaging in Adolescents." Paper presented to the Workshop on Mobile Communications: Understanding Users, Adoption and Design at the Conference on Human Factors in Computing Systems (CHI), Seattle, WA. Retrieved February 1, 2006, at http://www.cs.colorado.edu/~palen/chi_workshop/papers/EldridgeGrinter.pdf

Ellis, R., Goodyear, P., Prosser, M., and O'Hara, A. 2006. "How and What University Students Learn Through Online and Face-to-Face Discussion: Conceptions, Intentions and Approaches." *Journal of Computer Assisted Learning.* 22:4:244–56.

Ellison, N., Heino, R., and Gibbs, J. 2006. "Managing Impressions Online: Self-Presentation Processes in the Online Dating Environment." *Journal of Computer-Mediated Communication.* 11:2. Retrieved October 23, 2006, at http://jcmc.indiana.edu/vol11/issue2/ellison.html

Else, L. 2007. "Our Blackberries, Ourselves." *Utne Reader.* January/February: 10–11.

Engestrom, Y., Engestrom, R., and Karkkainen, M. 1995. "Polycontextuality and Boundary Crossing in Expert Cognition: Learning and Problem Solving in Complex Work Activities." *Learning and Instruction.* 5:4:319–36.

Erikson, E. H. 1993. *Childhood and Society.* New York: Norton.

Etzioni, A., and Etzioni, O. 1999. "Face-to-Face and Computer-Mediated Communities, A Comparative Analysis." *Information Society.* 15:4:241–49.

Fallows, D. 2006. "Growing Numbers Search the Web Just for Fun." *Pew Internet and American Life Project.* Retrieved November 10, 2006, at http://www.pewinternet.org/PPF/r/175/report_display.asp

Faraj, S., and Sproull, L. 2000. "Coordinating Expertise in Software Development Teams." *Management Science.* 46:12:1554–68.

Farrar, K. M., Krcmar, M., and Nowak, K. 2006. "Contextual Features of Violent Video Games, Mental Models, and Aggression." *Journal of Communication.* 56:2:387–405.

Feenberg, A., and Bakardjieva, M. 2004. "Virtual Community: No 'Killer Implication.' " *New Media & Society.* 6:1:37–43.

Fehr, D. 2005. *Friendship Processes.* New York: Sage.

Ferlander, S., and Timms, D. 2006. "Bridging the Dual Digital Divide: A Local Net and an IT-Café in Sweden." *Information, Communication and Society.* 9:2:137–59.

Fernback, J. 2007. "Beyond the Diluted Community Concept: A Symbolic Interactionist Perspective on Online Social Relations." *New Media & Society.* 9:1:49–69.

Finn, J. 1999. "An Exploration of Helping Processes in an Online Self-Help Group Focusing on Issues of Disability." *Health Social Work.* 24:3:220–31.

Finn, M. 2005. "Gaming Goes Mobile: Issues and Implications." *Australian Journal of Emerging Technologies and Society.* 3:1. Retrieved March 7, 2006, at http://www.swin.edu.au/sbs/ajets/journal/issue4/abstract_104Finn.htm

Fisher, D., Brush, A. J., Gleave, E., and Smith, M. 2006. "Revisiting Whittaker & Sidner's 'Email Overload' Ten Years Later." *CSCW.* 309–12. Retrieved November 28, 2006, at http://research.microsoft.com/research/pubs/view.aspx?type=Publication&id=1678

Fisher, H. 2004. *Why We Love*. New York: Henry Holt.

Flanagin, A. J., and Metzger, M. J. 2001. "Internet Use in the Contemporary Media Environment." *Human Communication Research*. 27:1:153–81.

Flor, N. V., and Hutchins, E. L. 1991. "Analyzing Distributed Cognition in Software Teams: A Case Study of Team Programming During Perfective Software Maintenance." *Empirical Studies of Programmers—Fourth Workshop*. Norwood, NJ: Ablex.

Flora, C. 2007. "Hi-Tech Tethers." *Psychology Today*. January/February: 51–53.

Floridi, L. 2007. "A Look into the Future Impact of ICT on our Lives." *Information Society*. 23:1:59–64.

Foot, K. A., and Schneider, S. M. 2002. "Online Action in Campaign 2000: An Exploratory Analysis of the U.S. Political Web Sphere." *Journal of Broadcasting and Electronic Media*. 46:2:222–44.

Foot, K.A., and Schneider, S. M. 2004. "Online Structure for Civic Engagement in the Post-9/11 Web Sphere." *Electronic Journal of Communication*. 14:3–4.

Foot, K., Warnick, B., and Schneider, S. M. 2005. "Web-Based Memorializing after September 11: Toward a Conceptual Framework." *Journal of Computer-Mediated Communication*. 11:1:4. Retreived September 13, 2006, at http://jcmc.indiana.edu/vol11/issue1/foot.html

Fortunati, L. 2000. "The Mobile Phone: New Social Categories and Relations." In Ling, R., and Trane, K. (eds.), *The Social Consequences of Mobile Telephony: The Proceedings from a Conference About Society, Mobile Telephony and Children*. June 26. Retrieved October 16, 2006, at http://www.telenor.no/fou/prosjekter/Fremtidens_Brukere/seminarer/mobilpresentasjoner/Proceedings%20_FoU%20notat_.pdf

Fortunati, L. 2002. "Italy: Stereotypes, True and False." In Katz, J. E. and Aakhus, M. A. (eds.), *Perpetual Contact: Mobile Communication, Private Talk, Public Performance*. Cambridge: Cambridge University Press.

Fountain, C. 2005. "Finding a Job in the Internet Age." *Social Forces*. 83:3:1235–62.

Fowler, R. B. 1991. *The Dance with Community*. Lawrence, KS: University of Kansas Press.

Fox, K. 2001: "Evolution, Alienation and Gossip: The Role of Mobile Telecommunications in the 21st Century." *Social Issues Research Center*, Oxford. Retrieved October 16, 2006, at http://www.sirc.org/publik/gossip.shtml

Fox, S. 2004. "The New Imagined Community: Identifying and Exploring a Bidirectional Continuum Integrating Virtual and Physical Communities through the Community Embodiment Model (CEM)." *Journal of Communication Inquiry*. 28:1:47–62.

Friedman, E. 2005. "The Reality of Virtual Reality: The Internet and Gender Equality Advocacy in Latin America." *Latin American Politics and Society*. 47:3:1–34.

Fung, A. 2006. "Bridging Cyberlife and Real Life: A Study of Online Communities in Hong Kong." In Silver, D., and Massanari, A. (eds.), *Critical Cyberculture Studies*. New York: New York University Press.

Gadamer, H. 1989. *Truth and Method* (2nd rev. ed.). Translated by Weinsheimer, J., and Marshall, D. G. New York: Continuum.

Galal, I. 2003. "Online Dating in Egypt." *Global Media Network.* 2:3. Retrieved June 5, 2006, at http://lass.calumet.purdue.edu/cca/gmj/fa03/graduate fa03/gmj_fa03_graduateTOC.htm#top

Galston, W. 1999. "Does the Internet Strengthen Community?" Retrieved January 10, 2005, at http://www.puaf.umd.edu/IPPP/fall1999/internet_community.htm

Gardner, R. O. 2004. "The Portable Community: Mobility and Modernization in Bluegrass Festival Life." *Symbolic Interaction.* 27:2:155–78.

Garud, R. 1997. "On the Distinction Between Know-How, Know-Why and Know-What in Technological Systems." In Huff, A. S. and Walsh, J. P. (eds.), *Advances in Strategic Management.* Greenwich, CT: JAI Press.

Gasson, S. 2005. "The Dynamics of Sensemaking, Knowledge, and Expertise in Collaborative, Boundary-Spanning Design." *Journal of Computer-Mediated Communication.* 10:4:14. Retrieved on September 13, 2006, at http://jcmc.indiana.edu/vol10/issue4/gasson.html

Gee, J. P. 2001. "Identity as an Analytic Lens for Research in Education." In Secada, W. (ed.), *Review of Research in Education, 25.* Washington, DC: American Educational Research Association.

Gergen, K. J. 1991. *The Saturated Self: Dilemmas of Identity in Contemporary Life.* New York: Basic Books.

Gergen, K. J., Gergen, M. M., and Barton, W. H. 1973. "Deviance in the Dark." *Psychology Today.* 7:129–30.

Geser, H. 1998. "Yours Virtually Forever: Death Memorials and Remembrance Sites in the WWW." *Sociology in Switzerland Online Publications.* Retrieved January 10, 2005, at http://socio.ch/intcom/t_hgeser07.htm

Geser, H. 2004. "Towards a Sociological Theory of the Mobile Phone." Unpublished manuscript. Retrieved July 15, 2006, at http://socio.ch/mobile/t_geser1.pdf

Gibbs, J., Ellison, N., and Heino, R. 2006. "Self-Presentation in Online Personals: The Role of Anticipated Future Interaction, Self-Disclosure, and Perceived Success in Internet Dating." *Communication Research.* 33:2:152–77.

Giddens, A. 1984. *The Constitution of Society: Outline of the Theory of Structuration.* Cambridge, UK: Polity Press.

Giddens, A. 1990. *The Consequences of Modernity.* Palo Alto, CA: Stanford University Press.

Giddens, A. 1992. *The Transformation of Intimacy: Sexuality, Love, and Eroticism in Modern Societies.* Palo Alto, CA: Stanford University Press.

Giddens, A. (with Beck, U., and Lash, S.) 1994. *Reflexive Modernization: Politics, Tradition, and Aesthetics in the Modern Social Order.* Cambridge, UK: Polity Press.

Gillard, P. 1996. "Women and New Technologies: Information and Telecommunications Needs Research (SIMS)." Monash University, Australia. Retrieved November 27, 2006, at http://www.infotech.monash.edu.au/itnr/reports/womentch.html

Gioia, D. A., Thomas, J. B., Clark, S. M., and Chittipeddi, K. 1994. "Symbolism and Strategic Change in Academia: The Dynamics of Sensemaking and Influence." *Organization Science.* 5:3:363–83.

Gladwell, M. 2006. "Here's Why." *The New Yorker.* April 10:80–82.

Glaser, J., Dixit, J., and Green, D. P. 2002. "Studying Hate Crime with the Internet: What Makes Racists Advocate Racial Violence?" *Journal of Social Issues.* 58:1:175–92.

Glasser, T. L. 1982. "Play, Pleasure and the Value of Newsreading." *Communication Quarterly.* 30:2:101–107.

Glasser, T. L. 2000. "Play and the Power of News." *Journalism: Theory, Practice, Criticism.* 1:1:23–29.

Goffman, E. 1959. *The Presentation of Self in Everyday Life.* New York: Anchor.

Goffman, E. 1974. *Frame Analysis: An Essay on the Organization of Experience.* New York: Harper.

Goffman, E. 1983. "Presidential Address: The Interaction Order." *American Sociological Review.* 48:1–17.

Goggin, G., and Newell, C. 2003. *Digital Disability: The Social Construction of Disability in New Media.* Lanham, MD: Rowman and Littlefield.

Goleman, D. 2006. *Social Intelligence.* New York: Bantam.

Goode, E., and Ben-Yehuda, N. 1994. *Moral Panics: The Social Construction of Deviance.* Oxford: Blackwell.

Goodwin, J., Jasper, J. M., and Polletta, F. (Eds.) 2001. *Passionate Politics.* Chicago: University of Chicago Press.

Gould, P., and White, R. 1974. *Maps in Minds.* Baltimore: Penguin.

Granovetter, M. 1973. "The Strength of Weak Ties." *American Journal of Sociology.* 78:1360-80.

Gray, T., Liscano, R., Wellman, B., Quan-Haase, A., Radhakrishnan, K., and Cho, Y. 2003. "Context and Intent in Call Processing." *Proceedings of Feature Interactions in Telecommunications and Software Systems VII.* Amsterdam: IOS Press.

Graya, N., Klein, J., Noycec, P., Sesselberg, T., and Cantriw, J. 2005. "Health Information-Seeking Behaviour in Adolescence: The Place of the Internet." *Social Science and Medicine.* 60:7:1467–78.

Green, M., Hilken, J., Friedman, H., Grossman, K., Gasiewski, J., Adler, R., and Sabini, J. 2005. "Communication via Instant Messenger: Short- and Long-Term Effects." *Journal of Applied Social Psychology.* 35:3:445–62.

Gregg, M. 2006. "Feeling Ordinary: Blogging as Conversational Scholarship." *Continuum: Journal of Media & Cultural Studies.* 20:2:147–60.

Griffiths, M. 2001. "Observations and Implications for Internet Sex Addiction." *Journal of Sex Research.* 38:4. Retrieved January 5, 2007, at http://find articles.com/p/articles/mi_m2372/is_4_38/ai_84866951

Grinter, R. E., and Eldridge, M. A. 2001. "Y Do Tngrs Luv 2 Txt Msg? In Prinz, W., Jarke, M., Rogers, Y., Schmidt, K., and Wulf, V. (eds.), *Proceedings of the Seventh European Conference on Computer Supported Cooperative Work, 16–20 September 2001, Bonn, Germany.* Dordrecht, Netherlands: Kluwer Academic Publishers.

Gross, N., and Simmons, S. 2002. "Intimacy as a Double-Edged Phenomenon? An Empirical Test of Giddens." *Social Forces*. 81:2:531–55.

Grossman, L. 2006. "Time's Person of the Year: You." *Time*. December 13. Retrieved December 19, 2006, at http://www.time.com/time/magazine/article/0,9171,1569514,00.html

Grotevant, H. D. 1998. "Adolescent Development in Family Contexts." In Damon, W., and Eisenberg, N. (eds.), *Handbook of Child Psychology; Vol. 3: Social, Emotional, and Personality Development, 5th ed.* New York: Wiley.

Guillén, M., and Suárez, S. 2005. "Explaining the Global Digital Divide: Economic, Political and Sociological Drivers of Cross-National Internet Use." *Social Forces*. 84:2:681–708.

Guldberg, K., and Pilkington, R. 2006. "A Community of Practice Approach to the Development of Non-Traditional Learners Through Networked Learning." *Journal of Computer Assisted Learning*. 22:3:159–71.

Guo, B., Bricout, J., and Huang, J. 2005. "A Common Open Space or a Digital Divide? A Social Model Perspective on the Online Disability Community in China." *Disability and Society*. 20:1:49–66.

Habermas, J. 1989. *The Structural Transformation of the Public Sphere*. Translated by Burger, M. Cambridge, MA: MIT Press.

Haddon, L. 2000. "The Social Consequences of Mobile Telephony: Framing." Paper presented at the seminar Sosiale Konsekvenser av Mobiltelefoni organized by Telenor, Oslo, June 16. Retrieved on October 16, 2006, at http://www.telenor.no/fou/prosjekter/Fremtidens_Brukere/seminarer/mobilpresentasjoner/Proceedings%20_FoU%20notat_.pdf

Hafner, K. 2006. "Laptop Slides into Bed in Love Triangle." *The New York Times*. August 24:G:1.

Hall, S., Critcher, C., Jefferson, T., Clarke, J., and Roberts, B. 1978. *Policing the Crisis: Mugging, the State, and Law and Order*. London: Macmillan.

Hallowell, E. M. 2006. *CrazyBusy: Overstretched, Overbooked and About to Snap*. New York: Ballantine Books.

Hamman, R. 2005. "An Apology to the Internauts." Retrieved May 19, 2005, at www.cybersociology.com

Hampton, K. N. 2003. "Grieving for a Lost Network: Collective Action in a Wired Suburb." *Information Society*. 19:5:417–29.

Hampton, K., and Wellman, B. 2002. "Not So Global Village of Netville." In Wellman, B., and Haythornthwaite, C. (eds.), *The Internet in Everyday Life*. Malden, MA: Blackwell.

Handelman, D. 1976. "Play and Ritual: Complementary Frames of Meta-Communication." In Chapman, A. J. and Foot, H. (eds.), *It's a Funny Thing, Humour*. London: Pergamon.

Hands, J. 2006. "Civil Society, Cosmopolitics and the Net: The Legacy of 15 February 2003." *Information, Communication & Society*. 9:2:225–43.

Hannigan, S. L., and Reinitz, M. T. 2001. "A Demonstration and Comparison of Two Types of Inference-Based Memory Errors." *Journal of Experimental Psychology: Learning, Memory, & Cognition*. 27:931–40.

Hanus, J. 2006. "The Culture of Pornography Is Shaping Our Lives, for Better and for Worse." *Utne.* September/October: 58–60.

Hardey, M. 2002. "Life Beyond the Screen: Embodiment and Identity Through the Internet." *The Sociological Review.* 50:4:570–85.

Hargittai, E. 2002. "Second-Level Digital Divide: Differences in People's Online Skills." *First Monday.* 7:4. Retrieved August 4, 2006, at http://firstmonday.org/issues/issue7_4/hargittai/index.html

Hargittai, E., and Shafer, S. 2006. "Differences in Actual and Perceived Online Skills: The Role of Gender." *Social Science Quarterly.* 87:2:432–48.

Hartman, G. B. 1995. "Public Memory and Its Discontents." In Brown, M. (ed.), *The Uses of History.* Durham, NC: Duke University Press.

Hartmann, T., and Klimmt, C. 2006. "Gender and Computer Games: Exploring Females' Dislikes." *Journal of Computer-Mediated Communication.* 11:4:2. Retrieved October 2, 2006, at http://jcmc.indiana.edu/vol11/issue4/hartmann.html

Haythornthwaite, C. 2002. "Building Social Networks via Computer Networks: Creating and Sustaining Distributed Learning Communities." In Renninger, K. A., and Shumar, W. (eds.), *Building Virtual Communities: Learning and Change in Cyberspace.* Cambridge: Cambridge University Press.

Haythornthwaite, C. 2005. "Social Networks and Internet Connectivity Effects." *Information, Communication and Society.* 8:2:125–47.

Henderson, S., and Gilding, M. 2004. " 'I've Never Clicked This Much with Anyone in My Life': Trust and Hyperpersonal Communication in Online Friendships." *New Media & Society.* 6:4:487–506.

Heritage, J. 1984. *Garfinkel and Ethnomethodology.* Cambridge: Polity.

Herring, S., Job-Sluder, K., Scheckler, R., and Barab, S. 2002. "Searching for Safety Online: Managing 'Trolling' in a Feminist Forum." *Information Society.* 18:5:371–85.

Hertzum, M. 2004. "Small-Scale Classification Schemes: A Field Study of Requirements Engineering." *Computer Supported Cooperative Work (CSCW).* 13:1:35–62.

Hewitt, B. 2006. "MySpace Nation: The Controversy." *People Magazine.* June 5: 113–21.

Hewson, C., Yule, P., Laurent, D., and Vogel, C. 2003. *Internet Research Methods: A Practical Guide for the Social and Behavioral Sciences.* London: Sage.

Hian, L. B., Chuan, S. L., Trevor, T. M. K., and Detenber, B. H. 2004. "Getting to Know You: Exploring the Development of Relational Intimacy in Computer-Mediated Communication." *Journal of Computer-Mediated Communication.* 9:3. Retrieved September 24, 2006, at http://jcmc.indiana.edu/vol9/issue3/index.html

Hine, T. 2000. *The Rise and Fall of the American Teenager.* New York: Harper Perennial.

Hollander, P. 2004. "The Counterculture of the Heart." *Society.* 41:2:69–77.

Hossain, L., and Wigand, R. T. 2004. "ICT Enabled Virtual Collaboration Through Trust." *Journal of Computer-Mediated Communication.* 10:1:8. Retrieved September 24, 2006, at http://jcmc.indiana.edu/vol10/issue1/index.html

Howard, P. N., and Jones, S. 2003. *Society Online: The Internet in Context.* Thousand Oaks, CA: Sage.

Howard, P. N. 2007. "Testing the Leap-frog Hypothesis: The Impact of Existing Infrastructure and Telecommunications Policy on the Global Digital Divide." *Information, Communication & Society.* 10:2:133–57.

Hu, Y., Wood, J. F., Smith, V., and Westbrook, N. 2004. "Friendships Through IM: Examining the Relationship Between Instant Messaging and Intimacy." *Journal of Computer-Mediated Communication.* 10:1:6. Retrieved September 24, 2006, at http://jcmc.indiana.edu/vol10/issue1/index.html

Huffaker, D. A., and Calvert, S. L. 2005. "Gender, Identity, and Language Use in Teenage Blogs." *Journal of Computer-Mediated Communication.* 10:1. Retrieved October 23, 2006, at http://jcmc.indiana.edu/vol10/issue2/huffaker.html

Huizinga, J. [1938] 1950. *Homo Ludens: A Study of the Play Element in Culture.* Boston: Beacon Press.

Hulme, M., and Peters, S. 2001. "Me, My Phone and I: The Role of the Mobile Phone." Paper presented at the Conference of Human Factors in Computing Systems (CHI) 2001 Workshop: Mobile Communications: Understanding Users, Adoption and Design.

Hutchins, E. 1991. "The Social Organization of Distributed Cognition." In Resnick, L. B., Levine, J. M., and Teasley, S. D. (eds.), *Perspectives on Socially Shared Cognition.* Washington, DC: American Psychological Association.

Igarashi, T., Takai, J., and Yoshida, T. 2005. "Gender Differences in Social Network Development via Mobile Phone Text Messages: A Longitudinal Study." *Journal of Social and Personal Relationships.* 22:5:691–713.

Ishii, K. 2006. "Implications of Mobility: The Uses of Personal Communication Media in Everyday Life." *Journal of Communication.* 56:2:346–65.

Israel, B. 2006. "The Overconnecteds." *The New York Times.* November 5. Retrieved December 14, 2006, at http://www.nytimes.com/2006/11/05/education/edlife/connect.html?ex=1184817600&en=9f5470487c75965b&ei=5070

Ito, M., and Okabe, D. 2005. "Technosocial Situations: Emergent Structuring of Mobile E-mail Use." In Ito, M., Okabe, D., and Matsuda, M. (eds.), *Personal, Portable, Pedestrian: Mobile Phones in Japanese Life.* Cambridge, MA: MIT Press.

Ivory, J. 2006. "Still a Man's Game: Gender Representation in Online Reviews of Video Games." *Mass Communication & Society.* 9:1:103–14.

Jackendoff, R. 1994. *Patterns in the Mind.* New York: Basic Books.

Jackson, L. A., von Eye, A., Barbatsis, G., Biocca, F., Fitzgerald, H. E., and Zhao, Y. 2004. "The Impact of Internet Use on the Other Side of the Digital Divide." *Communications of the ACM.* 47:7:43–47.

James, W. [1890] 1983. *Principles of Psychology.* Cambridge, MA: Harvard University Press.

Jewkes, Y., and Andrews, C. 2005. "Policing the Filth: The Problems of Investigating Online Child Pornography in England and Wales." *Policing & Society.* 15:1:42–62.

Johnson, B., Lorenz, E., and Lundvall, B. A. 2002. "Why All This Fuss About Codified and Tacit Knowledge?" *Industrial and Corporate Change.* 11:2:245–62.

Johnson, M. K. 1983. "A Multiple-Entry, Modular Memory System." In Bower, G. (ed.), *The Psychology of Learning and Motivation: Advances in Research and Theory*. New York: Academic Press.

Johnson, S. 2005. "Are the Kids Alright After All?" *New Scientist*. 187:2506: 48–49.

Johnson, S. 2006. *Everything Bad Is Good for You*. New York: Riverhead Books.

Jones, B. 1999. "Books of Condolence." In Walter, T. (ed.), *The Mourning of Diana*. New York: Oxford University Press.

Jones, S. 2003. "Let the Games Begin: Gaming Technology and Entertainment Among College Students." *Pew Internet and American Life Project*. Retrieved June 5, 2006, at www.pewinternet.org

Jordan, T. 1999. *Cyberpower: The Culture and Politics of Cyberspace and the Internet*. London: Routledge.

Jordan, T. 2001. "Language and Libertarianism: The Politics of Cyberculture and the Culture of Cyberpolitics." *Sociological Review*. 49:1–17.

Jorgenson-Earp, C. R., and Lanzilotti, L. A. 1998. "Public Memory and Private Grief: The Construction of Shrines at the Site of Public Tragedy." *Quarterly Journal of Speech*. 84:2:150–70.

Juul, J. 2003. "The Game, the Player, the World: Looking for a Heart of Gameness." In Copier, M., and Raessens, J. (eds.), *Level Up: Digital Games Research Conference Proceedings*. Utrecht: Utrecht University. Retrieved October 2, 2006, at http://www.jesperjuul.net/text/gameplayerworld

Juul, J. 2004. "Introduction to Game Time." In Wardrip-Fruin, N., and Harrigan, P. (eds.), *First Person: New Media as Story, Performance and Game*. Cambridge, MA: MIT Press.

Juul, J. 2005. *Half Real: Video Games Between Real Rules and Fictional Worlds*. Cambridge, MA: MIT Press.

Kahn, R., and Kellner, D. 2004. "New Media and Internet Activism: From the 'Battle of Seattle' to Blogging." *New Media & Society*. 6:1:87–95.

Kang, H., Bagchi-sen, S., and Rao, H. R. 2005. "Internet Skeptics: An Analysis of Intermittent Users and Net Dropouts." *IEEE Technology and Society Magazine*. 24:2:26–31.

Karbo, K. 2006. "Friendship: The Laws of Attraction." *Psychology Today*. November/December: 91–95.

Katz, J., and Aackhus, M. (eds.) 2002. *Perpetual Contact: Mobile Communication, Private Talk, Public Performance*. Cambridge: Cambridge University Press.

Katz, J., and Sugiyama, S. 2006. "Mobile Phones as Fashion Statements: Evidence from Student Surveys in the US and Japan." *New Media & Society*. 8:2:321–37.

Katz, J. E., and Rice, R. E. 2002. *Social Consequences of Internet Use*. Cambridge, MA: MIT Press.

Kauffman, R., and Wood, C. 2006. "Doing Their Bidding: An Empirical Examination of Factors That Affect a Buyer's Utility in Internet Auctions." *Information Technology and Management*. 7:3:171–90.

Kavada, A. 2005. "Exploring the Role of the Internet in the 'Movement for Alternative Globalization': The Case of the Paris 2003 European Social Forum."

Westminster Papers in Communication and Culture. 2:1:72–95. Retrieved September 20, 2006, at http://www.wmin.ac.uk/mad/pdf/Kavada.pdf

Kavanaugh, A., and Patterson, S. 2002. "The Impact of Computer Networks on Social Capital and Community Involvement in Blacksburg." In Wellman, B., and Haythornthwaite, C. (eds.), *The Internet in Everyday Life*. Malden, MA: Blackwell.

Kavanaugh, A., Reese, D., Carroll, J., and Rosson, M. 2005. "Weak Ties in Networked Communities." *Information Society*. 21:2:119–31.

Kayahara, J., and Wellman, B. 2007. "Searching for Culture—High and Low." *Journal of Computer-Mediated Communication*. 12:3. Retrieved June 3, 2007, at http://jcmc.indiana.edu/vol12/issue3/kayahara.html

Kelly, K. 2005. "We Are the Web." *Wired*. 13:08. Retrieved March 4, 2006, at http://www.wired.com/wired/archive/13.08/tech.html?pg=5&topic=tech&topic_set

Kendall, L. 2002. *Hanging Out in the Virtual Pub: Masculinities and Relationships Online*. Berkeley: University of California Press.

Keskin-Kozat, B. 2004. "Book Review: Liquid Love." *Contemporary Sociology*. 33:4:494–95.

Kharif, O. 2006. "Social Networking Goes Mobile." *Business Week Online*. May 31. Retrieved July 10, 2006, at www.businessweek.com/technology/content/May2006/tc20060530_170086.htm

Khazan, O. 2006. "Lost in an Online Fantasy World." *WashingtonPost.com*. August 18. Retrieved November 15, 2006, at http://www.washingtonpost.com/wp-dyn/content/article/2006/08/17/AR2006081700625.html?referrer=email&referrer=email&referrer=email

Kibby, M. 1997. "Babes on the Web: Sex, Identity and the Home Page." *Media International Australia*. 84:39–45.

Kidd, D. 2003. "Indymedia.org: A New Communications Commons." In McCaughey, M., and Ayers, M. D. (eds.), *Cyberactivism: Online Activism in Theory and Practice*. New York: Routledge.

Klastrup, L., and Tosca, S. 2004. "Transmedial Worlds—Rethinking Cyberworld Design." Presented at Proceedings International Conference on Cyberworlds 2004, IEEEE Computer Society, Los Alamitos, CA. Retrieved December 9, 2006, at http://www.itu.dk/people/klastrup/klastruptosca_transworlds.pdf

Kling, R., and Courtright, C. 2003. "Group Behavior and Learning in Electronic Forums: A Sociotechnical Approach." *Information Society*. 19:3:221–36.

Kawamoto, K. 2003. "Compassion Knows No Border: The Research of Patricia Radin." *Interface: The Journal of Education, Community and Values*. December. 3:9. Retrieved October 16, 2006, at http://bcis.pacificu.edu/journal/2003/09/kawamoto.php

Kornet, A. 1997. "The Truth About Lying." *Psychology Today*. May/June: 52–57.

Koyuncu, C., and Bhattacharya, G. 2004. "The Impacts of Quickness, Price, Payment Risk, and Delivery Issues on Online Shopping." *Journal of Socio-Economics*. 33:2:241–52.

Krueger, B. 2005. "Government Surveillance and Political Participation on the Internet." *Social Science Computer Review.* 23:4:439–52.

Kuhn, P., and Skuterud, M. 2004. "Internet Job Search and Unemployment Durations." *The American Economic Review.* 94:1:218–32.

Kulikova, S. V., and Perimutter, D. D. "Blogging Down the Dictator?" *International Communication Gazette.* 69:1:29–50.

Kumar, R., Liben-Nowell, D., Novak, J., Raghavan, P., and Tomkins, A. 2005. "Geographic Routing in Social Networks." *Proceedings of the National Academy of Science.* 102:33:11623–28.

Kuran, T., and McCaffery, E. J. 2004. "Expanding Discrimination Research: Beyond Ethnicity and to the Web." *Social Science Quarterly.* 85:3:713–30.

Lamb, B. 2004. "Wide Open Spaces: Wikis, Ready or Not. *Educause Review. September/October.* 39:5:36–48. Retrieved December 9, 2006, at https://www.educause.edu/pub/er/erm04/erm0452.asp

Langman, L. 2005. "From Virtual Public Spheres to Global Justice: A Critical Theory of Internetworked Social Movements." *Sociological Theory.* 23:1:42–74.

LaRose, R., Eastin, M. S., and Gregg, J. 2001. "Reformulating the Internet Paradox: Social Cognitive Explanations of Internet Use and Depression." *Journal of Online Behavior.* 1:2. Retrieved October 2, 2006, at http://www.behavior.net/JOB/v1n1/paradox.html

Lave, J., and Wenger, E. 1991. *Situated Learning: Legitimate Peripheral Participation.* Cambridge: Cambridge University Press.

Lawson, H., and Leck, K. 2006. "Dynamics of Internet Dating." *Social Science Computer Review.* 24:2:189–208.

Lea, M., and Spears, R. 1991. "Computer-Mediated Communication, De-Individuation, and Group Decision Making." *International Journal of Man-Machine Studies.* 34:283–301.

Lea, M., and Spears, R. 1995. "Love at First Byte? Building Personal Relationships Over Computer Networks." In Wood, J. T., and Duck, S. (eds.), *Understudied Relationships: Off the Beaten Track.* Thousand Oaks, CA: Sage.

Leach, E. 1976. *Culture and Communication.* Cambridge: Cambridge University Press.

Lee, H. 2005. "Behavorial Strategies for Dealing with Flaming in an Online Forum." *Sociological Quarterly.* 46:2:385–403.

Lee, H. 2006. "Privacy, Publicity, and Accountability of Self-Presentation in an Online Discussion Group." *Sociological Inquiry.* 76:1:1–22.

Legge, K. 2005. "Mobile Nation—Upwardly Mobile." *The Australian Magazine.* May 28. Retrieved March 7, 2006, at http://blogs.doctorcolossus.com/technology/?p=24

Leibowitz, J. 2001. *Knowledge Management: Learning from Knowledge Engineering.* Boca Raton, FL: CRC Press.

Lemann, N. 2006. "Amateur Hour." *The New Yorker.* August 7/14: 44–49.

Lenhart, A., and Madden, M. 2006. "Teen Content Creators and Consumers." *Pew Internet and American Life Project.* Retrieved October 3, 2006, at http://www.pewinternet.org/PPF/r/166/report_display.asp

Lenhart, A., and Fox, S. 2006. "Bloggers: A Portrait of the Internet's New Story-tellers." *Pew Internet and American Life Project*. Retrieved August 4, 2006, at http://www.pewinternet.org/report_display.asp?r=186

Lenhart, A., Madden, M., and Hitlin, P. 2005. "Teens and Technology." *Pew Internet and American Life Project*. Retrieved August 4, 2006, at http://www.pewinternet.org/PPF/r/162/report_display.asp

Lenhart, A., Rainie, L., and Lewis, O. 2001. "Teenage Life Online: The Rise of the Instant-Message Generation and the Internet's Impact on Friendships and Family Relationships." *Pew Internet and American Life Project*. Retrieved August 1, 2005, at http://www.pewinternet.org/PPF/r/36/report_display.asp

Levinson, P. 2004. *Cellphone: The Story of the World's Most Mobile Medium and How It Has Transformed Everything!* New York: Palgrave Macmillan.

Lewin, K. 1936. *Principles of Topological Psychology*. New York: McGraw-Hill.

Licoppe, C. 2004. " 'Connected' Presence: The Emergence of a New Repertoire for Managing Social Relationships in a Changing Communication Technoscape." *Environment and Planning D: Society and Space*. 22:1:135–56.

Licoppe, C., and Heurtin, J.-P. 2002. "Jean-Philippe France: Preserving the Image." In Katz, J. E. and Aakhus, M. A. (eds.), *Perpetual Contact: Mobile Communication, Private Talk, Public Performance*. Cambridge: Cambridge University Press.

Lin, D. 2006. "Sissies Online: Taiwanese Male Queers Performing Sissinesses in Cyberspaces." *Inter-Asia Cultural Studies*. 7:2:270–88.

Lin, N. 1999. "Building a Network Theory of Social Capital." *Connections*. 22:1:28–51. Retrieved October 30, 2006, at http://www.insna.org/Connections-Web/Volume22-1/V22(1)-28-51.pdf

Lindsay, S., Smith, S., Bell, F., and Bellaby, P. 2007. "Tackling the Digital Divide: Exploring the Impact of ICT on Managing Heart Conditions in a Deprived Area." *Information, Communication & Society*. 10:1:95–114.

Linenthal, E. T. 2001. *The Unfinished Bombing: Oklahoma City in American Memory*. Oxford: Oxford University Press.

Ling, J. 2007. "MySpace Mobile: Portable Communities Empowering Users." Presentation at O'Reilly ETel Emerging Telephony Conference, San Francisco, CA. February 27. Retrieved March 5, 2007, at http://conferences.oreillynet.com/cs/etel2007/view/e_sess/11773

Ling, R. 2004a. *The Mobile Connection*. San Francisco, CA: Morgan Kaufman.

Ling, R. 2004b. " 'I Have a Free Telephone So I Don't Bother to Send SMS, I Call': The Gendered Use of SMS Among Adults in Intact and Divorced Families" in Hflich, J. (ed.), *Qualitative Analysis of Mobile Communication*. J. Universitt Erfurt. Retrieved July 2, 2006, at http://www.richardling.com/papers/2004_Gendered_use_of_SMS.pdf

Ling, R. 2005a. "Flexible Coordination in the Nomos: Stress, Emotional Maintenance and Coordination via the Mobile Telephone in Intact Families." In Kavoori, A., and Arceneaux, N. (eds.), *Cultural Dialectics and the Cell Phone: Essays in Social Transformation*. New York: Peter Lang.

Ling, R. 2005b. "The Socio-Linguistics of SMS: An Analysis of SMS Use by a Random Sample of Norwegians." In Ling, R., and Pederson, P. (eds.), *Mobile Communications: Renegotiation of the Social Sphere*. London: Springer.

Ling, R., and Yttri, B. 2002. "Hypercoordination via Mobile Phones in Norway." In Katz, J. E., and Aakhus, M. A. (eds.), *Perpetual Contact: Mobile Communication, Private Talk, Public Performance*. Cambridge: Cambridge University Press.

Livingstone, S., Bober, M. and Helsper, E. 2005. "Active Participation or Just More Information?" *Information, Communication & Society*. 8:3:287–314.

Loftus, E. 1996. *Eyewitness Testimony*. Cambridge, MA: Harvard University Press.

Lombard, M. and Ditton, T. 1997. "At the Heart of It All: The Concept of Presence." *Journal of Computer-Mediated Communication*. 3:2. Retrieved November 1, 2006, at http://jcmc.indiana.edu/vol3/issue2/lombard.html

Lomrantz, T. 2006. "The Curse of Internet Addiction." *The New York Daily News* November 26: 35.

Losh, S. 2004. "Gender, Educational, and Occupational Digital Gaps." *Social Science Computer Review*. 22:2:152–66.

Lyon, D. 2004. "Globalizing Surveillance: Comparative and Sociological Perspectives." *International Sociology*. 19:2:135–49.

MacKinnon, G., and Williams, P. 2006. "Models for Integrating Technology in Higher Education." *Journal of College Science Teaching*. 35:7:22–25.

Madden, M. 2006. "Internet Penetration and Impact." *Pew Internet and American Life Project*. Retrieved November 16, 2006, at http://201.21.232.102/PPF/r/182/report_display.asp

Madden, M., and Lenhart, A. 2006. "Online Dating." *Pew Internet and American Life Project*. Retrieved October 3, 2006, at http://www.pewinternet.org/PPF/r/177/report_display.asp

Madge, C., and O'Connor, H. 2006. "Parenting Gone Wired: Empowerment of New Mothers on the Internet?" *Social & Cultural Geography*. 7:2:199–220.

Magid, L. 2001. "Europe Children Cell Use Ahead of U.S.—For Good, Bad." *The Mercury News*. December 13: 3.

Magid, L. 2007. "Global Positioning by Cellphone." *The New York Times*. July 19. Retrieved July 19, 2007, at http://www.nytimes.com/2007/07/19/technology/circuits/19basics.html?_r=1&8dpc&oref=slogin

Mann, C., and Stewart, F. 2000. *Internet Communication and Qualitative Research*. London: Sage.

Mari-kiose, P. 2004. "How Democracies Lose Small Wars: State, Society, and the Failures of France in Algeria, Israel in Lebanon, and the United States in Vietnam." *American Journal of Sociology*. 110:2:535–37.

Markus, M. L., and Bjorn-Andersen, N. 1987. "Power Over Users: Its Exercise by System Professionals." *Communications of the ACM*. 30:6:498–504.

Martin, S. P., and Robinson, J. P. 2007. "The Income Digital Divide: Trends and Predictions for Levels of Internet Use." *Social Problems*. 54:1:1–22.

Martini, E. 2003. "Public Histories, Private Memories?: Cybermemorials and the Future of Public History." Unpublished manuscript, University of Maryland.

Mason-Schrock, D. 1996. "Transsexuals' Narrative Construction of the 'True Self.' " *Social Psychology Quarterly*. 59:3:176–92.

Matsuba, M. 2006. "Searching for Self and Relationships Online." *CyberPsychology and Behavior.* 9:3:275–84.

McCaughey, M., and Ayers, M. D. (eds.) 2003. *Cyberactivism: Online Activism in Theory and Practice*. New York: Routledge.

McCullagh, D., and Mills, E. 2006. "Feds Take Porn Fight to Google." *CnetNews.com*. January 19. Retrieved December 4, 2006, at www.news.com/Feds_take_porn_fight_to-Google/2100-1030_3-6028701.html

McCullagh, D., and Broache, A. 2006. "FBI Taps Cell Phone Mic as Eavesdropping Tool." *CnetNews.com*. December 1. Retrieved December 4, 2006, at http://news.com.com/FBI+taps+cell+phone+mic+as+eavesdropping+tool/2100-1029_3-6140191.html?tag=sas.email

McDaniel, S. 2002. "Information and Communication Technologies: Bugs in the Generational Ointment." *Canadian Journal of Sociology*. 27:4:535–47.

McGowan, K. 2004. "Addiction: Pay Attention." *Psychology Today*. November/December. Retrieved December 10, 2006, at http://www.psychologytoday.com/articles/pto-20041111-000001.html

McKenna, K. Y. A. 2003. "Social Identity and the Self on the Internet." Paper presented at the Computer-Supported Social Interaction Conference, Miami University, April.

McKenna, K. Y. A., and Bargh, J. A. 1998. "Coming Out in the Age of the Internet: Identity 'Demarginalization' through Virtual Group Participation." *Journal of Personality and Social Psychology*. 75:681.

McKenna, K. Y. A., Green, A. S., and Gleason, M. 2002. "Relationship Formation on the Internet: What's the Big Attraction?" *Journal of Social Issues*. 58:1:9–31.

McMillan, S., and Morrison, M. 2006. "Coming of Age with the Internet: A Qualitative Exploration of How the Internet Has Become an Integral Part of Young People's Lives." *New Media & Society*. 8:1:73–95.

McQuaid, R., Lindsay, C., and Greig, M. 2004. " 'Reconnecting the Unemployed: Information and Communication Technology and Services for Jobseekers in Rural Areas." *Information, Communication & Society*. 7:3:364–88.

Mead, G. H. 1934. *Mind, Self, and Society*. Chicago: University of Chicago Press.

Meho, L. I. 2006. "E-Mail Interviewing in Qualitative Research: A Methodological Discussion." *Journal of the American Society for Information Science and Technology*. 57:10:1284–95.

Mehra, B., and Merkel, C. 2004. "The Internet for Empowerment of Minority and Marginalized Users." *New Media & Society*. 6:6:781–802.

Merton, R. K. 1957. *Social Theory and Social Structure*. New York: The Free Press.

Mesch, G., and Talmud, I. 2006. "The Quality of Online and Offline Relationships: The Role of Multiplexity and Duration of Social Relationships." *Information Society*. 22:3:137–48.

Metzger, M. J. 2007. "Communication Privacy Management in Electronic Commerce." *Journal of Computer-Mediated Communication*. 12:2. Retrieved July 10, 2007, at http://jcmc.indiana.edu/vol12/issue2/metzger.html

Meyrowitz, J. 1984. *No Sense of Place.* Oxford: Oxford University Press.

Mikula, M. 2003. "Virtual Landscapes of Memory." *Information, Communication & Society.* 6:2:169–87.

Mitra, A. 1997. "Virtual Commonality: Looking for India on the Internet." In Jones, S. G. (ed.), *Virtual Culture: Identity and Communication in Cybersociety.* London: Sage Publications.

Mitra, A. 2004. "Voices of the Marginalized on the Internet: Examples from a Website for Women of South Asia." *Journal of Communication.* 54:3:492–510.

Mitra, A. 2005. "Creating Immigrant Identities in Cybernetic Space: Examples from a Non-Resident Indian Website." *Media, Culture and Society.* 27:3:371–90.

Miyaki, Y. 2005. "Keitai Use Among Japanese Elementary and Junior High School Students." In Ito, M., Okabe, D., and Matsuda, M. (eds.), *Personal, Portable, Pedestrian: Mobile Phones in Japanese Life.* Cambridge, MA: MIT Press.

Miyata, K., Boase, J., Wellman, B. and Ikeda, K. 2005. "The Mobile-izing Japanese: Connecting to the Internet by PC and Webphone in Yamanashi." In Ito, M., Okabe, D., and Matsuda, M. (eds.), *Personal, Portable, Pedestrian: Mobile Phones in Japanese Life.* Cambridge, MA: MIT Press.

Monk, A., Carroll, J., Parker, S., and Blythe, M. 2004. "Why Are Mobile Phones Annoying?" *Behaviour and Information Technology,* 23:1:33–41.

Moore, R. 2004. "Beyond Our Control? Confronting the Limits of Our Legal System in the Age of Cyberspace." *International Social Science Review.* 79:3/4: 161–62.

Morahan-Martin, J. 2005. "Internet Abuse: Addiction? Disorder? Symptom? Alternative Explanations?" *Social Science Computer Review.* 23:1:39–48.

Morahan-Martin, J., and Schumacher, P. 2003. "Loneliness and Social Uses of the Internet." *Computers in Human Behavior.* 19:6:659–71.

Moreland, R. L. 1999. "Transactive Memory: Learning Who Knows What in Work Groups and Organizations." In Thompson, L., Levine, J. M., and Messick, D. M. (eds.), *Shared Cognition in Organizations: The Management of Knowledge.* Mahwah, NJ: Lawrence Erlbaum Associates.

Mulcahy, J. K. 1997. "Role Playing Characters and the Self." Retrieved August 20, 2003, at http://beyond3sigma.loki.ws/anthro.html

Mulkay, M. 1988. *On Humour: Its Nature and Its Place in Modern Society.* Cambridge, UK: Polity.

Murray, S. L., Holmes, J. G., and Griffin, D. W. 1996. "The Self-Fulfilling Nature of Positive Illusions in Romantic Relationships: Love Is Not Blind, but Prescient." *Journal of Personality and Social Psychology.* 71:1155–80.

Nakajima, I., Himeno, K., and Yoshii, H. 1999. "Diffusion of Cellular Phones and PHS and Its Social Meanings." *Joho Tsuushin Gakkai-shi.* 16:3:79–92.

Nakamura, L. 1999. "Race in/for Cyberspace: Identity Tourism and Racial Passing on the Internet." In Vitanza, V. J. (ed.), *CyberReader.* Boston, MA: Allyn and Bacon.

Nastri, J., Pena, J., and Hancock, J. T. 2006. "The Construction of Away Messages: A Speech Act Analysis." *Journal of Computer-Mediated Communication.* 11:4:7. Retrieved on October 16, 2006, at http://jcmc.indiana.edu/vol11/issue4/nastri.html

National Children's Home. 2005. "Putting U in the Picture: Mobile Bullying Survey 2005." London. Retrieved January 24, 2006, from http://www.nch.org.uk/uploads/documents/Mobile_bullying_%20report.pdf

Neimark, J. 1995. "It's Magical. It's Malleable. It's Memory." *Psychology Today*. 28:1:44–49, 80–85.

Newsweek Online. 2002. "Every Day Is a Gift, Isn't It?" November 16. Retrieved October 13, 2006, at http://www.msnbc.msn.com/id/3070155/site/newsweek/

Nicolini, D., Gherardi, S., and Yanow, D. (eds.). 2003. *Knowing in Organizations: A Practice-Based Approach*. New York: M. E. Sharpe.

Nippert-Eng, C. 1995. *Home and Work*. Chicago: University of Chicago Press.

Nippert-Eng, C. 2006. "Secrets and Secrecy." Chapter in unpublished manuscript, *Islands of Privacy*.

Nordlund, R., and Bartholet, J. 2001. "The Web's Dark Secret." *Newsweek*. March 1: 44–51.

Nonaka, I., and Konno, N. 1998. "The Concept of 'Ba': Building Foundation for Knowledge Creation." *California Management Review*. 40:3:40–54.

Nonnecke, B., and Preece, J. 2000. "Lurker Demographics: Counting the Silent." Paper presented at the CHI, April, Amsterdam.

Nonnecke, B., and Preece, J. 2001. "Why Lurkers Lurk." Proceedings of 7 Americas Conference on Information Systems (AMCIS).

Norrick, N. R. 1993. *Conversational Joking: Humor in Everyday Talk*. Bloomington: Indiana University Press.

Nowak, K. L., Watt, J., and Walther, J. B. 2005. "The Influence of Synchrony and Sensory Modality on the Person Perception Process in Computer-Mediated Groups." *Journal of Computer-Mediated Communication*. 10:3:3. Retrieved on September 24, 2006, at http://jcmc.indiana.edu/vol10/issue3/nowak.html

O'Brien, J. 1999. "Writing in the Body: Gender (Re)production in Online Interaction." In Smith, M. A., and Kollock, P. (eds.), *Communities in Cyberspace*. New York: Routledge.

O'Harrow, R. 2005. *No Place to Hide*. New York: Simon and Schuster.

Okabe, D. 2004. "Emergent Social Practices, Situations and Relations through Everyday Camera Phone Use." Paper presented at the 2004 International Conference on Mobile Communication. October 18–19. Seoul, Korea.

Oksman, V., and Rautianen, P. 2002. " 'Perhaps It Is a Body Part': How the Mobile Phone Became An Organic Part of the Everyday Lives of Finnish Children and Adolescents." In Katz, J. (ed.), *Machines That Become Us*. New Brunswick: Transaction Publishers.

Oldenburg, R. 1999. *The Great Good Place: Cafés, Coffee Shops, Community Centers, Beauty Parlors, General Stores, Bars, Hangouts, and How They Get You Through the Day*. New York: Marlowe and Company.

Page, S. 2006. "NSA Secret Database Report Triggers Fierce Debate in Washington." *USA Today*. May 11. Retrived December 11, 2006 from www.usatoday.com/news/Washington/2006-05-11-nsa-reax_x.htm

Palen, L., Salzman, M., and Youngs, E. 2000. "Going Wireless: Behavior & Practice of New Mobile Phone Users." Paper presented at the Computer Supported

Cooperative Work Conference by the Association for Computing Machinery, Philadelphia, PA.

Palen, L., Hiltz, S. R., and Liu, S. B. 2007. "Online Forums Supporting Grassroots Participation in Emergency Preparedness and Response." *Communications of the ACM.* 50:3:54–58.

Palmer, J. 1994. *Taking Humour Seriously.* London: Routledge.

Pan, B., Hembrooke, H., Joachims, T., Lorigo, L., Gay, G., and Granka, L. 2007. "In Google We Trust: Users' Decisions on Rank, Position, and Relevance." *Journal of Computer-Mediated Communication.* 12:3. Retrieved July 10, 2007, at http://jcmc.indiana.edu/vol12/issue3/pan.html

Paul, P. 2005. *Pornified.* New York: Times Books.

Pavlik, J. 2000. "The Impact of Technology on Journalism." *Journalism Studies.* 1:2:229–37.

Pecchioni, L. L., and Sparks, L. 2007. "Health Information Sources of Individuals with Cancer and Their Family Members." *Health Communication.* 21:2:143–51.

Persichitte, K. A., Young, S., and Tharp, D. D. 1997. "Conducting Research on the Internet: Strategies for Electronic Interviewing." In Proceedings of Selected Research and Development Presentations at the 1997 National Convention of the Association for Educational Communications and Technology. February 14–18, Albuquerque, NM. Washington, DC: Association for Educational Communications and Technology.

Peter, J., and Valkenburg, P. 2006a. "Research Note: Individual Differences in Perceptions of Internet Communication." *European Journal of Communication.* 21:2:213–26.

Peter, J., and Valkenburg, P. 2006b. "Adolescents' Exposure to Sexually Explicit Material on the Internet." *Communication Research.* 33:2:178–204.

Peter, J., and Valkenburg, P. 2007. "Who Looks for Casual Dates on the Internet? A Test of the Compensation and the Recreation Hypothesis." *New Media & Society.* 9:3:455–74.

Petronio, S. 2002. *Boundaries of Privacy, Dialetics of Disclosure.* Albany, NY: State University of New York Press.

Pickard, V. 2006. "United Yet Autonomous: Indymedia and the Struggle to Sustain a Radical Democratic Network." *Media, Culture and Society.* 28:3: 315–36.

Pisello, T. 2007. "E-Mail Overload Costs Organizations Over $5,000 per User per Year." *IT Business Edge.* March 22. Retrieved April 10, 2007, at http://www. itbusinessedge.com/blogs/tom/?p=1

Pitts, V. 2004. "Illness and Internet Empowerment: Writing and Reading Breast Cancer in Cyberspace." *Health: An Interdisciplinary Journal for the Social Study of Health, Illness and Medicine.* 8:1:33–59.

Plant, S. 1997. *Zeroes and Ones: Digital Women and the New Technoculture.* New York: Doubleday.

Polanyi, M. 1958. *Personal Knowledge: Towards a Post-Critical Philosophy.* Chicago: University of Chicago Press.

Porter, C. E. 2004. "A Typology of Virtual Communities: A Multi-Disciplinary Foundation for Future Research." *Journal of Computer-Mediated Communication.* 10:1. Retrieved December 7, 2004, at www.ascusc.org/jcmc/vol10/issue1/porter.html

Portes, A. 1998. "Social Capital: Its Origins and Applications in Modern Sociology." *Annual Review of Sociology.* 22:1–24.

Poster, M. 1995. *The Second Media Age.* Cambridge: Polity Press.

Potter, B. 2006. "Wireless Hotspot: Petri Dish of Wireless Security." *Communications of the ACM.* 49:6:51–56.

Prager, K. J. 1995. *The Psychology of Intimacy.* New York: Guilford.

Preece, J. 1998. "Empathic Communities: Reaching Out Across the Web." *Interactions.* 2:2:32–43.

Preece, J. 1999. "Empathy Online." *Virtual Reality.* 4:1–11.

Preece, J. 2000. *Online Communities: Designing Usability, Supporting Sociability.* New York: John Wiley and Sons.

Prusak, L. 2001. "Where Did Knowledge Management Come From?" *IBM Systems Journal.* 40:4:1002–1007.

Psychology Today. 2004. "Cupid's Comeuppance." September/October. Retrieved September 5, 2006, at http://www.psychologytoday.com/articles/pto-20040921-000001.html

Psychology Today. 2007. "Texting Gr8 4U." *Psychology Today.* January/February: 14.

Puro, Jukka-Pekka. 2002. "Finland: A Mobile Culture." In Katz, J. E., and Aakhus, M. A. (eds.), *Perpetual Contact: Mobile Communication, Private Talk, Public Performance.* Cambridge: Cambridge University Press.

Putnam, R. 1995. "Bowling Alone, Revisited." *The Responsive Community*, Spring: 18–33.

Quan-Haase, A., Cothrel, J., and Wellman, B. 2005. "Instant Messaging for Collaboration: A Case Study of a High-Tech Firm." *Journal of Computer-Mediated Communication.* 10:4:13. Retrieved on October 16, 2006, at http://jcmc.indiana.edu/vol10/issue4/quan-haase.html

Quan-Haase, A., and Wellman, B. 2002. "Understanding the Use of Communication Tools for Ad-Hoc Problem-Solving in Mid-Size Organizations." Paper presented at the Popular Culture Association and American Culture Association Conference: Electronic Culture and Communications Forum, Toronto, ON, March 13–16.

Raab, C., and Mason, D. 2004. "Privacy, Surveillance, Trust and Regulation." *Information, Communication and Society.* 7:1:89–91.

Radcliffe, A. M., Lumley, M. A., Kendall, J., Stevenso, J. K., and Beltran, J. 2007. "Written Emotional Disclosure: Testing Whether Social Disclosure Matters." *Journal of Social and Clinical Psychology.* 26:4:362–84.

Radin, P. 2003. "Online Medical Communication Among Peers: The Net and Alternatives to Traditional Journalism." In Kawamoto, K. (ed.), *Digital Journalism: Emerging Media and the Changing Horizons of Journalism.* Lanham, MD: Rowman and Littlefield.

Radin, P. 2006. " 'To Me, It's My Life': Medical Communication, Trust, and Activism in Cyberspace." *Social Science and Medicine*. 62:3:591–601.

Ragin, C. 1987. *The Comparative Method: Moving Beyond Qualitative and Quantitative Strategies*. Berkeley: University of California Press.

Ragin, C. 1994. *Constructing Social Research: The Unity and Diversity of Method*. Thousand Oaks, CA: Pine Forge Press.

Rakow, L., and Navarro, V. 1993. "Remote Mothering and the Parallel Shift: Women Meet the Cellular Telephone." *Critical Studies in Mass Communication*. 20:3:144–57.

Rainie, L. 2005. "Search Engine Use Shoots Up in the Past Year." *Pew Internet and American Life Project*. Retrieved November 14, 2006, at http://www.pewinternet.org/PPF/r/167/report_display.asp

Rainie, L. 2006. "Digital Natives Invade the Workplace." *Pew Internet and American Life Project*. Retrieved December 6, 2006, at http://pewresearch.org/obdeck/?ObDeckID=70

Ramirez, A., Walther, J. B., Burgoon, J. K., and Sunnafrank, M. 2002. "Information-Seeking Strategies, Uncertainty, and Computer-Mediated Communication: Toward a Conceptual Model." *Human Communication Research*. 28:2:213–28.

Reid, D., and Reid, F. 2004. "Insights into the Social and Psychological Effects of SMS Text Messaging." Unpublished manuscript. February. Retrieved October 2, 2006, at http://mail.cse.edu/exchange/scanzonieri/Inbox/ref%20115%20-%20journ%20art%20%22Reid%20and%20Reid%22.EML/1_multipart_xF8FF_2_SocialEffectsOfTextMessaging.pdf/C58EA28C-18C0-4a97-9AF2-036E93DDAFB3/SocialEffectsOfTextMessaging.pdf?attach=1

Reid, E. 1999. "Hierarchy and Power: Social Control in Cyberspace." In Smith, M. A., and Kollock, P. (eds.), *Communities in Cyberspace*. New York: Routledge.

Renninger, K. A., and Shumar, W. 2002. "Community Building with and for Teachers at the Math Forum." In Renninger, K. A., and Shumar, W. (eds.), *Building Virtual Communities: Learning and Change in Cyberspace*. Cambridge: Cambridge University Press.

Reynolds, G. 2006. *An Army of Davids: How Markets and Technology Empower Ordinary People to Beat Big Media, Big Government and Other Goliaths*. Salisbury, Wiltshire, UK: Nelson.

Rheingold, H. 1993. *The Virtual Community*. Reading, MA: Addison-Wesley.

Rheingold, H. 2002. *The Virtual Community, revised edition*. Cambridge, MA: MIT Press.

Rheingold, H. 2006. *Smart Mobs: The New Social Revolution*. New York: Perseus.

Richardson, J., and Swan, K. 2003. "Examining Social Presence in Online Courses in Relation to Students' Perceived Learning and Satisfaction." *Journal of Asynchronous Learning Networks*. 7:1. Retrieved July 1, 2005, at www.aln.org/publications/jaln/v7n1/pdf/v7n1)richardson.pdf

Ricoeur, P. 1976. *Interpretation Theory: Discourse and the Surplus of Meaning*. Fort Worth: Texas Christian University Press.

Ridings, C., and Gefen, D. 2004. "Virtual Community Attraction: Why People Hang Out Online." *Journal of Computer-Mediated Communication.* 10:1:4. Retrieved October 2, 2006, at http://jcmc.indiana.edu/vol10/issue1/index.html

Riva, G. and Davide, F. 2003. *Being There: Concepts, Effects and Measurement of User Presence in Synthetic Environments.* Amsterdam: IOS Press.

Roberts, P., and Videl, L. A. 1999. "Perpetual Care in Cyberspace: A Portrait of Memorials on the Web." *Omega: The Journal of Death and Dying.* 40:4:521–45.

Robbins, K., and Turner, M. A. 2002. "United States: Popular, Pragmatic and Problematic." In Katz, J. E., and Aakhus, M. (eds.), *Perpetual Contact: Mobile Communication, Private Talk, Public Performance.* Cambridge, UK: Cambridge University Press.

Roloff, M. E., and Solomon, D. H. 1989. "Sex Typing, Sports Interests, and Relational Harmony." In Wenner, I. (ed.), *Media, Sports, and Society.* Newbury Park, CA: Sage.

Rogers, C. 1951. *The True Self: Client-Centered Therapy.* Boston: Houghton-Mifflin.

Rosen, J. 2004. "Your Blog or Mine?" *The New York Times Magazine.* December 19. Retrieved September 10, 2006, at http://www.nytimes.com/2004/12/19/magazine/19PHENOM.html?ex=1261198800&en=0f68277267a43d84&ei=5090&partner=rssuserland

Rosson, M. B. 1999. "Get By With A Little Help From My Cyber-Friends: Sharing Stories of Good and Bad Times on the Web." *Journal of Computer-Mediated Communication.* 4:4. Retrieved on September 24, 2006, at http://jcmc.indiana.edu/vol4/issue4/index.html

Roxborough, I. 2004. "Review Essay: Thinking About War." *Sociological Forum.* 19:3:505–28.

Ryle, G. [1949] 1984. *The Concept of Mind.* Chicago: University of Chicago Press.

Sæther, B. 1998. "Retroduction: An Alternative Research Strategy?" *Business Strategy and the Environment.* 7:4:245–49.

Sakkopoulos, E., Lytras, M., and Tsakalidis, A. 2006. "Adaptive Mobile Web Services Facilitate Communication and Learning Internet Technologies." *IEEE Transactions on Education.* 49:2:208–15.

Sanders, C., Field, T. M., Diego, M., and Kaplan, M. 2000. "The Relationship of Internet Use to Depression and Social Isolation." *Adolescence.* 35:138:237–42.

Sandvig, C. 2006. "The Internet at Play: Child Users of Public Internet Connections." *Journal of Computer-Mediated Communication.* 11:4:3. Retrieved November 1, 2006, at http://jcmc.indiana.edu/vol11/issue4/sandvig.html

Sassen, S. 2004. "Local Actors in Global Politics." *Current Sociology.* 52:4: 649–70.

Satchell, C., Zic, J., Singh, S. 2005. "Creating the Ideal Digital Self." Paper presented at Pervasive Image, Capturing and Sharing Workshop, Ubicomp. September. Tokyo, Japan.

Schatzki, T. R., Knorr Cetina, K., and Von Savigny, E. (eds.) 2001. *The Practice Turn in Contemporary Theory*. New York: Routledge.

Schmidt, K. 1997. *Of Maps and Scripts: Proceedings of GROUP 97 ACM SIG, Distributed Group Work*. Phoenix, AZ: University of Phoenix.

Schopler, J. H., Abell, M. D., and Galinsky, M. 1998. "Technology-Based Groups: A Review and Conceptual Framework for Practice." *Social Work*. 43:254–67.

Schneider, S. M., and Dougherty, M. 2003. "Strategic Co-production and Content Appropriation: Press Materials on Candidate Web Sites in the 2002 U.S. Election." Paper presented at the meeting of the International Communication Association, San Diego.

Schramm, W. 1954. "How Communication Works." In Schramm, W. (ed.), *The Process and Effects of Mass Communication*. Urbana, IL: University of Illinois Press.

Schrock, D., Holden, D., and Reid, L. 2004. "Creating Emotional Resonance: Interpersonal Emotion Work and Motivational Framing in a Transgender Community." *Social Problems*. 51:1:61.

Schutz, A. 1951. "Making Music Together: A Study in Social Relationship." *Social Research*. 18:76–97.

Schutz, A. 1962. *Collected Papers*. The Hague: Martinus Nijhoff.

Schutz, A. 1973. "On Multiple Realities." In Natanson, M. (ed.), *Collected Papers of Alfred Schutz, Vol. 1*. The Hague: Martinus Nijhoff.

Schwandt, T. 1998. "Constructivist, Interpretivist Approaches to Human Inquiry." In Denzin, N. K., and Lincoln, Y. S. (eds.), *The Landscape of Qualitative Research: Theories and Issues*. Thousand Oaks, CA: Sage.

Schwartz, J. 2002. "The Nation: Case-Sensitive Crusader; Who Owns the Internet? You and I Do." *The New York Times*. December 29. Retrieved May 10, 2005, at query.nytimes.com

ScienceDaily.com. 2005. "Internet Dating Much More Successful Than Once Thought." February 14. Retrieved May 19, 2005, at www.sciencedaily.com/releases/2005/02/050218125144.htm

Shah, D., Cho, J., Eveland Jr., W., and Kwak, N. 2005. "Information and Expression in a Digital Age: Modeling Internet Effects on Civic Participation." *Communication Research*. 32:5:531–65.

Shanken, A. M. 2002. "Planning Memory: Living Memorials in the United States During World War II." *The Art Bulletin*. 84:1:130–48.

Sheller, M. 2004. "Mobile Publics: Beyond the Network Perspective." *Environment and Planning D: Society and Space*. 22:1:39–52.

Shibutani, T. 1955. "Reference Groups as Perspectives." *American Journal of Sociology*. 60:562–69.

Short, J., Williams, E., and Christie, B. 1976. *The Social Psychology of Telecommunications*. London: John Wiley and Sons.

Siegl, E., and Foot, K. A. 2004. "Expression in the Post-September 11th Web Sphere." *Electronic Journal of Communication*. 14:1–2.

Silver, D. 2000. "Looking Backwards, Looking Forward: Cyberculture Studies 1990–2000." Resource Center for Cyberculture Studies. Retrieved March 15, 2006, at http://rccs.usfca.edu/intro.asp

Silver, D. 2003. "Epilogue: Current Directions and Future Questions." In Mc-Caughey, M., and Ayers, M. D. (eds.), *Cyberactivism: Online Activism in Theory and Practice*. New York: Routledge.

Silver, D., and Massanari, A. (eds.) 2006. *Critical Cyberculture Studies*. New York: New York University Press.

Simmel, G. 1898. "The Persistence of Social Groups." *American Journal of Sociology*. 3:5:662–98.

Simmel, G. 1906. "The Sociology of Secrecy and of Secret Societies." *American Journal of Sociology* 11:441–98.

Simmel, G. [1908] 1962. *Conflict and the Web of Group Affiliations*. New York: Free Press.

Simmel, G. 1964. *The Sociology of Georg Simmel*. Translated by K. H. Wolff. Glencoe, IL: Free Press.

Skogan, W. G. 2004. "Policing Contingencies." *American Journal of Sociology*. 110:1:259–61.

Skovholt, K., and Svennevig, J. (2006). "Email Copies in Workplace Interaction." *Journal of Computer-Mediated Communication*. 12:1. Retrieved December 1, 2006, at http://jcmc.indiana.edu/vol12/issue1/skovholt.html

Smircich, L., and Morgan, G. 1982. "Leadership: The Management of Meaning." *Journal of Applied Behavioral Science*. 18:3:257–73.

Solove, D. J. 2004. *The Digital Person: Technology and Privacy in the Information Age*. New York: New York University Press.

Spears, R., Postmes, T., Lea, M., and Wolbert, A. 2002. "When Are Net Effect Gross Products? The Power of Influence and the Influence of Power in Computer-Mediated Communication." *Journal of Social Issues*. 58:1:91–108.

Spitzberg, B. H. 2006. "Preliminary Development of a Model and Measure of Computer-Mediated Communication (CMC) Competence." *Journal of Computer-Mediated Communication*. 11:2:12. Retrieved on October 2, 2006, at http://jcmc.indiana.edu/vol11/issue2/spitzberg.html

Sproull, L., and Kiesler, S. 1993. "Computers, Networks and Work." In Harasim, L. (ed.), *Global Networks: Computers and International Communication*. Cambridge, MA: MIT Press.

Star, S. L. 1989. "The Structure of Ill-Structured Solutions: Boundary Objects and Heterogeneous Distributed Problem Solving." In Gasser, L., and Huhns, M. N. (eds.), *Distributed Artificial Intelligence, Vol. II*. San Mateo, CA: Morgan Kaufmann.

Steinkuehler, C., and Williams, D. 2006. "Where Everybody Knows Your (Screen) Name: Online Games as 'Third Places.' " *Journal of Computer-Mediated Communication*. 11:4:1. Retrieved on October 2, 2006, at http://jcmc.indiana.edu/vol11/issue4/steinkuehler.html

Stephenson, W. 1964. "The Ludenic Theory of Newsreading." *Journalism Quarterly*. 41:3:367–74.

Stephenson, W. 1967. *The Play Theory of Mass Communication*. Chicago: University of Chicago Press.

Sterling, B. 1992. *The Hacker Crackdown: Law and Disorder on the Electronic Frontier*. London: Viking.

Stoller, E., and Longino, C. 2004. "Portable Communities: The Impact of Chain Migration and Ethnicity on Informal Networks of Retired Sunbelt Migrants." Presentation for Gerontological Society of America, Washington, DC. November.

Stone, A. R. 1996. *The War of Desire and Technology at the Close of the Mechanical Age.* Cambridge, MA: MIT Press.

Stone, B., and Richtel, M. 2007. "Social Networking Leaves Confines of the Computer." *The New York Times.* April 30. Retrieved June 1, 2007, at http://www.nytimes.com/2007/04/30/technology/30social.html?ex=118 4817600&en=336cab50ebdce812&ei=5070

Strauss, A. L, and Corbin, J. 1990. *Basics of Qualitative Research: Grounded Theory Procedures and Techniques.* Newbury Park, CA: Sage.

Strauss, A. L., and Corbin, J. 1998. *Basics of Qualitative Research: Techniques and Procedures for Developing Grounded Theory, 2nd Edition.* Newbury Park, CA: Sage.

Strogatz, S. 2003. *Sync: The Emerging Science of Spontaneous Order.* New York: Hyperion.

Suchman, L. 1987. *Plans and Situated Action.* Cambridge, MA: Cambridge University Press.

Suchman, L. 1996. "Constituting Shared Workspaces." In Engestrom, Y., and Middleton, D. (eds.), *Cognition and Communication at Work.* New York: Cambridge University Press.

Suellentrop, C. 2007. "Playing with Our Heads." *Utne.* January/February: 58–63.

Suhail, K., and Bargees, Z. 2006. "Effects of Excessive Internet Use on Undergraduate Students in Pakistan." *CyberPsychology and Behavior.* 9:3:297–301.

Suler, J. 2004. "The Online Disinhibition Effect." *CyberPsychology and Behavior.* 7:321–26. Retrieved December 2, 2006, at http://www.rider.edu/suler/psycyber/disinhibit.html

Sveningsson, M. 2002. "Cyberlove: Creating Romantic Relationships on the Net." In Fornas, J., et al. (eds.), *Digital Borderlands: Cultural Studies of Identity and Interactivity on the Internet.* New York: Peter Lang.

Tannen, D. 1991. *You Just Don't Understand: Men and Women in Conversation.* London: Virago.

Tanner, E. 2001. "Chilean Conversations: Internet Forum Participants Debate Augusto Pinochet's Detention." *Journal of Communication.* 51:2:383–404.

Tanner, K. J. 2005. "Emotion, Gender and the Sustainability of Communities." *Journal of Community Informatics.* 1:2:121–30.

Taylor, J. R., and Every, E. J. 2000. *The Emergent Organization: Communication as Its Site and Surface.* Mahwah, NJ: Lawrence Erlbaum Associates.

Thomas, A. 2006. " 'MSN Was the Next Big Thing after Beanie Babies': Children's Virtual Experiences as an Interface to Their Identities and Their Everyday Lives." *E-Learning.* 3:2:1–17.

Thomas, W. I., and Thomas, D. S. 1928. *The Child in America.* New York: Knopf.

Thompson, K. 1998. *Moral Panics.* London: Routledge.

Thoreau, E. 2006. "Ouch!: An Examination of the Self-Representation of Disabled People on the Internet." *Journal of Computer-Mediated Communication.* 11:2:3. Retrieved September 13, 2006, at http://jcmc.indiana.edu/vol11/issue2/thoreau.html

Thornburgh, N. 2006. "Parents for Poker." *Time.com.* September 25. Retrieved November 17, 2006, at http://www.time.com/time/magazine/article/0,9171,1538649,00.html

Thurlow, C., Lengel, L., and Tomic, A. 2004. *Computer Mediated Communication: Social Interaction and the Internet.* London: Sage.

Tilly, C. 2006. *Why? What Happens When People Give Reasons . . . and Why?* Princeton, NJ: Princeton University Press.

Tonn, B. 2004. "MyEmpowerNet.gov: A Proposal to Enhance Policy E-Participation." *Social Science Computer Review.* 22:3:335–46.

Tonnies, F. [1887] 1963. *Community and Society.* New York: Harper.

Tugend, A. 2006. "Debating the Age of Consent for That First Cellphone." *The New York Times.* September 30. Retrieved February 10, 2007, at http://www.nytimes.com/2006/09/30/business/30shortcuts.html?ex=1317268800&en=3c76201cda995c16&ei=5088&partner=rssnyt&emc=rss

Tugend, A. 2007. "Too Busy to Notice You're Too Busy." *The New York Times.* March 31. Retrieved July 10, 2007, at http://www.nytimes.com/2007/03/31/business/31shortcuts.html?ex=1184990400&en=74effac79493de01&ei=5070

Turkle, S. 1995. *Life on the Screen: Identity in the Age of the Internet.* New York: Simon and Schuster.

Turow, J., and Hennessy, M. 2007. "Internet Privacy and Institutional Trust: Insights from a National Survey." *New Media & Society.* 9:2:300–18.

UN News Center. 2001. "UN Report on Bridging Digital Divide." Retrieved January 5, 2005, at http://www.un.org/NEWS/

Urbina, I. 2006. "In Online Mourning, Don't Speak Ill of the Dead." *The New York Times.* Retrieved on December 18, 2006, at http://www.nytimes.com/2006/11/05/us/05memorial.html?ex=1320382800&en=ab8534741367a5fd&ei=5088&partner=rssnyt&emc=rss

Utz, S. 2000. "Social Information Processing in MUDs: The Development of Friendships in Virtual Worlds." *Journal of Online Behavior.* 1:1. Retrieved October 16, 2003, at http://www.behavior.net/job/v1n1/utz.html

Valkenburg, P., Schouten, A., and Peter, J. 2005. "Adolescents' Identity Experiments on the Internet." *New Media & Society.* 7:3:383–402.

Van Lear, C. A., Sheehan, M., Withers, L. A., and Walker, R. A. 2005. "AA Online: The Enactment of Supportive Computer Mediated Communication." *Western Journal of Communication.* 69:1:5–27.

van Dijk, J. A. G. M. 2005. *The Deepening Divide: Inequality in the Information Society.* Thousand Oaks, CA: Sage.

van 't Hooft, M., and Kelly, J. 2004. "Macro or Micro: Teaching Fifth-Grade Economics Using Handheld Computers." *Social Education.* 68:2:165–68.

Venkatesh, M. 2003. "The Community Network Lifecycle: A Framework for Research and Action." *Information Society.* 19:339–47.

Venolia, G., Gupta, A., Cadiz, J. J., and Dabbish, L. 2001. "Supporting Email Workflow." *Microsoft Technical Report*. MSR-TR-2001-88.

Vilhjalmsson, H. H., and Cassell, J. 1998. "BodyChat: Autonomous Communicative Behaviors in Avatars." *Proceedings of the Second International Conference on Autonomous Agents*. 269–76. Minneapolis, MN.

Virnoche, M. E., and Marx, G. T. 1997. "Only Connect—E.M. Forster in an Age of Electronic Communication: Computer-Mediated Association and Community Networks." *Sociological Inquiry*. 67:1:85–100.

Viseu, A., Clement, A., and Aspinall, J. 2004. "Situating Privacy Online." *Information, Communication and Society*. 7:1:92–114.

Vroman, A. 2007. "Australian Young People's Participatory Practices and Internet Use." *Information, Communication and Society*. 10:1:48–68.

Wagner, G., Pischner, R., and Haisken-DeNew, J. P. 2002. "The Changing Digital Divide in Germany." In Wellman, B., and Haythornthwaite, C. (eds.), *The Internet in Everyday Life*. Malden, MA: Blackwell.

Wallace, P. 1999. *The Psychology of the Internet*. Cambridge: Cambridge University Press.

Walter, T. (ed.) 1999. *The Mourning for Diana*. New York: Oxford University Press.

Walther, J. B. 1996. "Computer-Mediated Communication: Impersonal, Interpersonal, and Hyperpersonal Interaction." *Communication Research*. 23:3–43.

Walther, J. 1997. "Group and Interpersonal Effects in International Computer-Mediated Collaboration." *Human Communication Research*. 23:342–69.

Walther, J., and D'Addario, K. P. 2001. "The Impacts of Emoticons on Message Interpretation in Computer-Mediated Communication." *Social Science Computer Review*. 19:324–47.

Walther, J., and Parks, M. R. 2002. "Cues Filtered Out, Cues Filtered In: Computer-Mediated Communication and Relationships." In Knapp, M. L., and Daly, J. A. (eds.), *Handbook of Interpersonal Communication*. Thousand Oaks, CA: Sage.

Walther, J., and Tidwell, L. C. 1995. "Nonverbal Cues in Computer-Mediated Communication, and the Effect of Chronemics on Relational Communication." *Journal of Organizational Computing and Electronic Commerce*. 5:4:355–78.

Walther, J., Loh, T., and Granka, L. 2005. "Let Me Count the Ways: The Interchange of Verbal and Nonverbal Cues in Computer-Mediated and Face-to-Face Affinity." *Journal of Language and Social Psychology*. 24:1:36–65.

Walther, J., Slovacek, C., and Tidwell, L. C. 2001. "Is a Picture Worth a Thousand Words? Photographic Images in Long Term and Short Term Virtual Teams." *Communication Research*. 28:105–34.

Warren, R. L. 1978. *The Community in America*, 3rd ed. Boston: Houghton Mifflin.

Wasko, M. M., and Faraj, S. 2000. " 'It Is What One Does': Why People Participate and Help Others in Electronic Communities of Practice." *Journal of Strategic Information Systems*. 9:2–3:155–73.

Wei, R., and Lo, V. H. 2006. "Staying Connected While on the Move: Cell Phone Use and Social Connectedness." *New Media & Society.* 8:1:53–72.

Weick, K. E. 1995. *Sensemaking in Organizations.* Thousand Oaks, CA: Sage.

Weinberg, N., Schmale, J., Uken, J., and Wessel, L. 1996. "Online Help: Cancer Patients Participate in a Computer-Mediated Support Group." *Health and Social Work.* 27:1:24–29.

Wellman, B. 1979. "The Community Question." *American Journal of Sociology.* 84:1201–31.

Wellman, B. 1997. "An Electronic Group Is Virtually a Social Network." In Kiesler, S. (ed.), *Culture of the Internet.* Mahwah, NJ: Lawrence Erlbaum.

Wellman, B. 2001. "Physical Place and CyberPlace: The Rise of Personalized Networking." *International Journal of Urban and Regional Research.* 25:2:227–52.

Wellman, B. 2002. "Little Boxes, Glocalization, and Networked Individualism." In Tanabe, M., van den Besselaar, P., and Ishida, T. (eds.), *Digital Cities II: Computational and Sociological Approaches.* Berlin: Springer. Retrieved July 6, 2006, at http://www.chass.utoronto.ca/~wellman/publications/littleboxes/littlebox.PDF

Wellman, B., and Gulia, M. 1999. "Net Surfers Don't Ride Alone: Virtual Communities as Communities." In Smith, M. A., and Kollack, P. (eds.), *Communities in Cyberspace.* London: Routledge.

Wellman, B., and Hampton, K. 1999. "Living Networked On and Offline." *Contemporary Sociology.* 28:6:648–54.

Wellman, B., and Haythornthwaite, C. (eds.) 2002. *The Internet in Everyday Life.* Oxford: Blackwell.

Wellman, B., Salaff, J., Dimitrova, D., Garton, L., Gulia, M., and Haythornthwaite, C. 1996. "Computer Networks as Social Networks: Collaborative Work, Telework, and Virtual Community." *Annual Review of Sociology.* 22:1:213–38.

Wellman, B., Hogan, B., Berg, K., Boase, J., Carrasco, J., Côté, R., Kayahara, J., Kennedy, T. L. M., and Tran, P. 2006. "Connected Lives: The Project." In Purcell, P. (ed.), *Networked Neighbourhoods.* London: Springer.

Wenger, E. 1998. *Communities of Practice: Learning, Meaning and Identity.* New York: Cambridge University Press.

Weslander, E. 2006. "Virtual-Reality Crimes Present Literal Challenge for Real-life Police." *Lawrence Journal World.* November 12. Retrieved December 5, 2006, at http://www.people.ku.edu/~nbaym/

Whittaker, S., and Sidner, C. 1996. "Email Overload: Exploring Personal Information Management of Email." *Proceedings of CHI 96 Conference on Human Factors in Computing Systems.* New York: ACM Press.

Whittaker, S., Jones, Q., and Terveen, L. 2002. "Persistence and Conversation Stream Management: Conversation and Contact Management." In *Proceedings of HICCS'02.* New York: IEEE Press.

Whitty, M. 2005. "The Realness of Cybercheating: Men's and Women's Representations of Unfaithful Internet Relationships." *Social Science Computer Review.* 23:1:57–67.

Widdifield, R., and Grover, V. 1995. "Internet and the Implications of the Information Superhighway for Business." *Journal of Systems Management.* May/June: 16–21.

Wiggins, J. D., and Lederer, D. A. 1984. "Differential Antecedents of Infidelity in Marriage." *American Mental Health Counselors Association Journal.* 6:4:152–61.

Willard, N. 2007. *Cyber-Safe Kids, Cyber-Savvy Teens: Helping Young People to Use the Internet Safely and Responsibly.* San Francisco: Jossey-Bass.

Williams, D. 2006. "On and Off The 'Net: Scales for Social Capital in an Online Era. *Journal of Computer-Mediated Communication.* 11:2:11. Retrieved October 2, 2006, at http://jcmc.indiana.edu/vol11/issue2/williams.html

Williams, M. 2007. "Policing and Cybersociety: The Maturation of Regulation within an Online Community." *Policing and Society.* 17:1:59–82.

Williams, S., and Williams, L. 2005. "Space Invaders: The Negotiation of Teenage Boundaries Through the Mobile Phone." *Sociological Review.* 53:2: 314–31.

Williamson, K., Wright, S., Schauder, D., and Bow, A. 2001. "The Internet for the Blind and Visually Impaired." *Journal of Computer Mediated Communication.* 7:1 Retrieved October 23, 2006, at http://jcmc.indiana.edu/vol7/issue1/williamson.html

Willis, S., and Tranter, B. 2006. "Beyond the 'Digital Divide': Internet Diffusion and Inequality in Australia." *Journal of Sociology.* 42:1:43–59.

Wilson, B., and Atkinson, M. 2005. "Rave and Straightedge, The Virtual and the Real: Exploring Online and Offline Experiences in Canadian Youth Subcultures." *Youth and Society.* 36:3:276–311.

Wilson, T., and Tan, H. P. 2005 "Less Tangible Ways of Reading." *Information, Communication & Society.* 8:3:394–416.

Winnicott, D. W. 1971. *Playing and Reality.* London: Tavistock.

Wood, R. T., and Williams, R. J. 2007. "Problem Gambling on the Internet: Implications for Internet Gambling Policy in North America." *New Media & Society.* 9:3:520–42.

Yates, J., and Orlikowski, W. 2002. "Genre Systems: Structuring Interaction Through Communicative Norms." *The Journal of Business Communication.* 39:1:13–35.

Yee, N. 2006. *The Daedalus Project.* Retrieved July 17, 2006, from http://www.nickyee.com/daedalus/archives/001468.php

Young, K. S. 1999. "Cybersexual Addiction." Retrieved April 5, 2005, from http://www.netaddiction.com/cybersexual_addiction.htm

Young, K. S., Griffin-Shelly, E., Cooper, A., O'Mara, J., and Buchanan, J. 2000. "Online Infidelity: A New Dimension in Couple Relationships with Implications for Evaluation and Treatment." *Sexual Addiction and Compulsivity.* 7:1/2:59–74.

Young, S. 2006. "Student Views of Effective Online Teaching in Higher Education." *The American Journal of Distance Education.* 20:2:65–77.

Yu, H. 2006. "From Active Audience to Media Citizenship: The Case of Post-Mao China." *Social Semiotics.* 16:2:303–26.

Yurchisin, J., Watchravesringkan, K., and Brown McCabe, D. 2005. "An Exploration of Identity Re-Creation in the Context of Internet Dating." *Social Behavior and Personality: An International Journal.* 33:8:735–50.

Zack, M. H. 1999. "Managing Codified Knowledge." *Sloan Management Review.* 40:4:45–58.

Zerubavel, E. 1981. *Hidden Rhythms: Schedules and Calendars in Social Life.* Chicago: University of Chicago Press.

Zerubavel, E. 1991. *The Fine Line: Making Distinctions in Everyday Life.* New York: Free Press.

Zerubavel, E. 1993. "Horizons: On the Sociomental Foundations of Relevance." *Social Research.* 60:2:397–413.

Zerubavel, E. 2003. *Time Maps: Collective Memory and the Social Shape of the Past.* Chicago: University of Chicago Press.

Zhao, S. 2003. " 'Being There' and the Role of Presence Technology." In Riva, G., and Davide, F. (eds.), *Being There: Concepts, Effects and Measurement of User Presence in Synthetic Environments.* Amsterdam: IOS Press.

Zhao, S. 2006. "Do Internet Users Have More Social Ties? A Call for Differentiated Analyses of Internet Use." *Journal of Computer-Mediated Communication.* 11:3:8. Retrieved October 2, 2006, at http://jcmc.indiana.edu/vol11/issue3/zhao.html

Zimmerman, E. 2004. "Narrative, Interactivity, Play, and Games: Four Naughty Concepts in Need of Discipline." In Wardrip-Fruin, N., and Harrigan, P. (eds.), *First Person: New Media as Story, Performance and Game.* Cambridge, MA: MIT Press.

INDEX

Aackhus, M., 243n. 16, 245n. 1,
    245n. 16, 245n. 23, 246n. 29
Aarseth, E., 69, 240n. 38
Abell, M. L., 243n. 13, 249n. 36
Academia. *See* education; distance
    learning; youth and learning
Activism. *See* political participation
Addiction. *See* compulsive use
Adultery. *See* cybercheating
Agar, J., 246n. 29
Age and aging. *See* sociomental con-
    nections and age
Agres, C., 244n. 58, 252n. 7, 252n. 9
Alcohol use, 85
Alexander, B., 244n. 46
Alienation, 120, 157, 200
Altruism online, 53
"Always on" mode, 91, 115. *See also*
    ambient copresence
Amazon.com, 4
Ambient copresence, 39, 91, 115–140,
    157, 161. *See also* presence; prox-
    imity; "always-on" mode; away
    messages
Amichai-Hamburger, Y., 144, 242n.
    70, 244n. 38, 244n. 39, 247n. 6
An, Y. J., 248n. 20
Anderson, B., 8, 233n. 8, 234n. 15,
    234n. 2, 254n. 74
Anderson, J., 242n. 73
Andrews, C., 252n. 21
Anger. *See* moods
Anime, 28, 33, 38, 43, 77, 215

Anonymity, 35, 45–49, 52–54, 59, 76,
    138, 147, 166, 170, 175, 187,
    206
Anxiety and technological use, 55–62,
    67, 74, 97, 100, 113–114, 117,
    120–130, 136, 144–145, 156,
    169–173, 179, 186–190. *See also*
    email overload
AOL (America Online), 132
Armour, S., 246n. 48, 246n. 51
Art and works of art, 29, 69, 105,
    161, 231
Atkinson, M., 181, 251n. 112
Attention span. *See* continuous partial
    attention
Availability. *See* presence
Avatars, 4, 8, 31, 163, 178
Avivram, I., 242n. 70, 244n. 38
Away messages, 39, 91, 115–117,
    177, 219
Ayers, M. D., 196, 247n. 72, 253n.
    41, 253n. 43, 253n. 60

Bacon, P., 196, 253n. 50, 253n. 55,
    253n. 61
Bagchi-Sen, S., 252n. 15
Bakardjieva, M., 234n. 20
Baker, A., 235n. 10, 235n. 21, 238n.
    10, 238n. 17, 238n. 18, 238n.
    34, 238n. 35, 238n. 36, 239n.
    45, 241n. 52, 241n. 62, 241n.
    63, 241n. 64, 241n. 66, 243n.
    27, 243n. 28, 243n. 29, 243n.

Baker, A. (*continued*)
31, 244n. 35, 244n. 36, 244n.
37, 247n. 8, 247n. 9, 247n. 12,
249n. 33, 249n. 50, 250n. 56,
250n. 61, 250n. 72, 250n. 89,
251n. 91, 251n. 94, 251n. 97
Baker, P., 21, 26, 45, 49, 99, 239n.
58, 239n. 60, 252n. 32, 252n.
36
Barabasi, A. L., 235n. 26
Barak, A., 252n. 24
Bargees, Z., 242n. 70, 242n. 76
Bargh, J. A., 171, 199, 238n. 15,
238n. 21, 238n. 22, 250n. 60,
250n. 62, 253n. 67
Barlow, J. P., 29, 236n. 31
Barrett, C., 166, 175, 249n. 47,
250n. 80
Bartholet, J., 246n. 47
Barton, W. H., 238n. 27
Baseball, 36. *See also* sports
Bateson, G., 240n. 9
Bautsch, H., 246n. 34, 247n. 62,
247n. 7
Baym, N., 73, 74, 181, 207, 241n.
47, 241n. 48, 241n. 50, 241n.
52, 241n. 54, 241n. 55, 241n.
56, 248n. 20, 254n. 75, 254n.
10, 254n. 14
Beck, J., 68, 240n. 30, 240n. 35,
251n. 109
Beder, J., 55, 239n. 54
Bell, C., 233n. 12
Bennett, L., 244n. 49, 248n. 18
Ben-Yehuda, N., 252n. 36
Ben-Ze'ev, A., 100, 241n. 58, 241n.
62, 241n. 64, 243n. 28, 244n.
38
Berger, P., 40, 117, 141, 237n. 68,
237n. 69, 237n. 70, 245n. 14,
247n. 1
Bhattacharya, G., 244n. 58
Biegel, S., 252n. 23
Blogs and blogging, 4, 5, 8, 9, 20,
23, 30–38, 44–46, 57, 69, 82,
91, 96, 104–105, 116, 136–138,

142, 148, 163, 166–178, 191,
193, 195, 209, 211, 215–231
academic, 105, 138
and identity, 166–169, 175–178
as community memory project,
34–36
as mass emails, 96–97, 148–149
blog rings, 137
live blogging, 29, 142
political blogs, 191–196
Blood, R., 149, 166, 168, 176–177,
234n. 16, 247n. 11, 249n. 45,
249n. 46, 249n. 49, 250n. 82,
251n. 98
Boase, J., 11, 234n. 26, 243n. 8,
253n. 71
Bober, M., 248n. 16
Bodmer, C., 240n. 18
Bond, R., 242n. 70
Boredom and technology use, 66, 67,
69–73, 82, 83, 106, 122, 129,
209, 212, 215, 231
Bornstein, R., 80
Bourdieu, P., 233n. 13
Bowker, N., 250n. 65
boyd, d., 44, 162–163, 176, 187,
189, 234n. 16, 237n. 6, 242n.
3, 246n. 32, 247n. 9, 248n. 10,
249n. 22, 249n. 23, 249n. 25,
249n. 26, 249n. 29, 249n. 31,
250n. 83, 250n. 85, 252n. 27,
252n. 33, 252n. 34, 252n. 35
Brennan, F., 26, 235n. 19, 241n. 52,
243n. 13
Bretag, T., 241n. 52, 241n. 54
Bricout, J., 250n. 68
Brignall, T., 248n. 16
Broache, A., 246n. 50
Brockman, J., 249n. 24, 249n. 36,
249n. 41, 251n. 102
Broom, A., 239n. 53
Brown, J. S., 236n. 44
Brown, P., 241n. 52
Brunner, H., 241n. 50
Bryant, J. A., 243n. 9, 243n. 13,
249n. 26

Buerkle, H., 74, 239n. 49, 241n. 53, 242n. 70, 244n. 45
Bulletin board, online. *See* discussion board
Bullying, 187. *See also* harassment
Burkhlater, B., 250n. 55, 251n. 103
Burgoon, J. K., 238n. 14
Burnett, G., 34, 38, 74, 236n. 47, 237n. 61, 237n. 62, 239n. 49, 241n. 53, 242n. 70, 244n. 45
Burt, R. S., 233n. 13

Calhoun, C., 234n. 14, 253n. 72
Caller ID, 141, 143–144
Calvert, S. L., 234n. 16, 248n. 2, 248n. 10, 248n. 15, 249n. 30, 250n. 53, 251n. 90, 251n. 105, 251n. 114
Camcorders, 4, 17. *See also* video sharing; vidcast; vodcast
Cameras, camera phones. *See* photos
Campbell, S., 239n. 45, 243n. 27, 252n. 30, 252n. 31
CamWorld, 166–167
Carmichael, P., 236n. 49, 250n. 75
Cassel, B., 234n. 16, 242n. 70, 246n. 47
Cassidy, J., 250n. 79
Castells, M., 120, 239n. 45, 241n. 40, 241n. 44, 241n. 60, 242n. 82, 243n. 13, 243n. 16, 243n. 20, 245n. 16, 245n. 20, 245n. 22, 245n. 23, 246n. 29, 246n. 30, 246n. 32, 246n. 36, 246n. 39, 246n. 43, 246n. 44, 246n. 46, 246n. 52, 246n. 59, 246n. 59, 248n. 12, 248n. 13, 248n. 16, 248n. 19, 249n. 25, 249n. 32, 249n. 37, 249n. 52, 251n. 93
Castronova, J. R., 252n. 22
Caughey, J., 40, 233n. 3, 237n. 69, 244n. 62
Celebrities, 91, 174, 220
Cerulo, K., 29, 236n. 28, 236n. 30, 236n. 54, 237n. 7, 237n. 8, 238n. 13

Chat rooms, 8, 9, 20, 29, 55, 85, 92, 98, 109, 142, 152, 165, 211, 216–218, 228, 233n. 10
Chayko, M., 233n. 7, 234n. 13, 234n. 22, 234n. 24, 235n. 15, 237n. 70, 237n. 7, 241n. 64, 243n. 8, 244n. 41, 247n. 1, 253n. 71, 253n. 73, 254n. 77
and *Connecting*, 3, 207, 233n. 2, 233n. 10, 234n. 13, 234n. 18, 234n. 22, 234n. 23, 234n. 2, 235n. 3, 235n. 6, 235n. 7, 235n. 8, 235n. 9, 235n. 10, 235n. 15, 235n. 18, 235n. 20, 235n. 22, 236n. 27, 236n. 28, 236n. 32, 237n. 67, 237n. 74, 237n. 3, 237n. 8, 238n. 14, 239n. 48, 242n. 81, 242n. 1, 244n. 62, 245n. 15, 245n. 19, 246n. 30, 247n. 74, 247n. 3, 247n. 15, 251n. 107, 251n. 108, 251n. 114, 253n. 40, 253n. 72, 254n. 1, 254n. 13
Children. *See* youth
Cho, J., 252n. 14
Christensen, T. H., 237n. 66
Christian, A., 236n. 37
Christie, B., 237n. 57
Civic journalism. *See* blogs, political
Civil liberties, 132–133, 139, 186, 190, 198, 218
Clark, H., 26, 235n. 19, 241n. 52, 244n. 42
Class, social. *See* social status
Closeness. *See* intimacy
Cognitive infrastructure of society, 18, 41, 62, 97, 109, 198, 201
Cognitive resonance, 18, 25–31, 37, 74, 77, 81, 192, 201
Colleagues, 22, 44, 75, 114, 138, 145, 207, 230–231. *See also* work and workplace
Collective identity. *See* identity, collective
Collective memory, 18, 21, 31–38, 42, 74

Commenting and interactive feedback, 34–35, 38–39, 45, 54, 74, 107, 136, 144, 148–149, 168–169, 176

Commerce online, 43, 51, 56, 70, 78, 101, 107–110, 131, 186, 198, 218

Common ground, 23, 26, 50, 66, 74

Communities of the mind, 11

Complementary differentiation, 2

Compulsive use of technology, 64, 78–86, 99, 145, 147, 165, 174, 200, 209, 223, 225–226

Consumerism, 78, 84, 85, 131–132, 184

Continuous partial attention, 71, 116, 122, 134, 180

Control in technology use, issues related to, 59, 79, 87, 93, 128, 138, 141–158, 183–202

Convenience of technology, 13, 50, 87, 90, 93–97, 100, 107, 113, 123, 135, 139, 151, 152, 194, 201, 206

Cook, J., 236n. 44

Cooley, C., 234n. 2, 248n. 4, 248n. 6, 248n. 8

Cooper, A., 243n. 27, 243n. 28, 244n. 39

Cothrel, J., 242n. 3

Coughlin, K., 253n. 40

Courtright, C., 244n. 51

Cramer, K., 244n. 50

Credibility. See trust

Crime, 186–193, 198, 201. See also harassment

Cross, R., 247n. 17

Csikszentmihalyi, M., 68, 240n. 34, 240n. 36

Curtis, P., 235n. 10

Cuthbert, A. J., 244n. 42, 244n. 48, 244n. 51

Cybercheating, 76, 99–100

Cybersex, 29, 97–110, 202

D'addario, K., 238n. 18

Danet, B., 64, 240n. 9, 240n. 10, 240n. 11, 240n. 15, 240n. 18, 240n. 36

Danowitz, A. K., 252n. 9

Dating. See love and romance; flirting

Davide, F., 245n. 1

Davis, M., 40

Dayan, D., 236n. 28

Dean, J., 184, 189, 250n. 74, 252n. 6, 252n. 10, 253n. 37, 253n. 38

Death. See web memorials

Deception. See secrets

Depression. See mental health and illness

DeWolfe, C., 190

Digital divide, 161, 184–186, 191

Disability. See sociomental connections and disability

Disconnection. See isolation

Discussion boards, 8, 10, 20, 32, 57, 72, 102–107, 132, 139, 142, 169, 173, 174, 211, 215–231

Disinhibition, 48, 54, 59, 104, 166, 187. See also anonymity

Disruptions. See flaming, trolling

Distance learning, 32, 58, 101, 103–105, 107, 109, 185, 211, 215–231. See also education

Dittmar, H., 242n. 70

Ditton, T., 236n. 28, 237n. 58

Dixit, J., 250n. 75

Dodson, S., 242n. 5

Donath, J., 159, 248n. 5, 250n. 87, 250n. 88, 252n. 28

Douglas, J. Y., 240n. 34, 240n. 37

Downs, R. N., 235n. 15

Drentea, P., 239n. 47

Drugs and drug use, 28, 40, 78, 84–85, 186–187

Du, H. S., 244n. 47

Durkheim, E., 6, 18, 141, 158, 233n. 9, 235n. 3, 235n. 4, 235n. 5, 247n. 1, 248n. 26

Eastin, M. S., 239n. 55
eBay, 108
Economic issues. *See* commerce
Edberg, D., 244n. 58, 252n. 7
Education and technology, 69, 88, 101, 103–105, 164–165, 184–186, 197–202. *See also* distance learning; youth and learning; information
Electronic Frontier Foundation, 193
Eldridge, M., 233n. 6, 243n. 20, 247n. 13, 249n. 27
Electronic interview and interviewing, 5, 205–213. *See also* online research; qualitative research
Ellis, R., 244n. 47
Ellison, N., 243n. 31, 248n. 1, 250n. 87, 251n. 94, 251n. 95, 251n. 97
Else, L., 237n. 4, 239n. 62, 242n. 79, 245n. 17, 246n. 29, 246n. 32, 246n. 45, 247n. 61, 249n. 25
Email, 3–5, 8, 9, 11, 20, 43, 46, 56, 57, 63, 71, 75–76, 88, 90, 94, 96, 97, 106–107, 115, 126, 129, 132, 135, 139, 142–158, 163, 173, 176–178, 194, 196, 205–213, 215–231
email overload, 126–127
filters, 144, 164
frequency of accessing, 142
"round robins" or email chains. *See* text circles
Emergencies, 95, 113, 120–130, 132, 185–186, 209
Emoticons, 46
Epinions, 108
Erikson, E. H., 249n. 42, 251n. 106
Etzioni, A., 233n. 11, 236n. 44, 236n. 45
Etzioni, O., 233n. 11, 236n.44, 236n. 45

Facebook, 4, 8, 44, 58, 163
Fallows, D., 239n. 1, 240n. 39

Families and family issues, 3, 9, 12, 17, 18–20, 30, 43–45, 50, 52–56, 57, 70, 75, 77, 81, 88, 89–96, 117–140, 143, 146–158, 167, 169, 173, 178–180, 200–201, 215–231
Faraj, S., 239n. 2
Farrar, K. M., 240n. 21
Fatigue. *See* physical discomfort and fatigue
Feenberg, A., 234n. 20
Fehr, D., 237n. 5
Ferlander, S., 252n. 14
Fernback, J., 233n. 12, 234n. 13
Ferrara, K., 241n. 50
Fessenden, J., 244n. 49, 248n. 18
Fiction and fanfiction, 33, 40–41, 66, 79, 189
Finn, J., 239n. 49
Fisher, D., 235n. 25, 246n. 33
Fitzsimmons, G. M., 238n. 15
Flaming, 59–60, 187–188
Flanagin, A. J., 236n. 45, 245n. 15
Flirting, 63–64, 73–86, 97–110
Flora, C., 242n. 79, 242n. 3, 245n. 3, 245n. 10, 247n. 8
Floridi, L., 251n. 111
Flow, 68
Flynn, N., 132
Foot, K. A., 236n. 52, 236n. 53, 237n. 56
Fortunati, L., 91, 133–134, 156, 242n. 3, 243n. 18, 246n. 34, 246n. 56, 246n. 57, 248n. 18
Fountain, C., 244n. 57
Fowler, R. B., 233n. 11
Fox, K., 74, 120, 157–158, 236n. 28, 241n. 57, 241n. 59, 241n. 60, 241n. 61, 242n. 3, 243n. 7, 245n. 3, 245n. 5, 245n. 21, 248n. 25, 248n. 27
Fox, S., 32, 234n. 16, 236n. 38, 236n. 39, 247n. 10, 249n. 45, 249n. 48, 250n. 80
Frick, T., 248n. 20

Friedman, E., 250n. 62
Friendship, friends, and "friending,"
    8–9, 12, 17–23, 29, 36, 38, 42,
    44–51, 55–57, 59, 65–66, 70–76,
    80, 82, 87–110, 114–119, 129,
    131, 134–139, 143, 146–158,
    162–169, 172, 174, 177–180,
    183, 209, 215–231
Fun. *See* play
Fung, A., 66, 240n. 21, 240n. 22

Gadamer, H., 240n. 5
Galal, I., 238n. 28, 238n. 32, 238n.
    33, 238n. 34, 239n. 45, 239n.
    56, 243n. 31, 243n. 33, 244n.
    36, 245n. 13, 249n. 44, 251n.
    96, 251n. 97
Galinsky, M. J., 243n. 13, 249n. 36
Galston, W., 243n. 17
Gambling, 28, 43, 61, 78, 79, 84, 85,
    229
Games and gaming, 18, 20, 43, 63–
    69, 71, 77, 78–80, 84, 128, 134,
    211, 217, 229. *See also* massive
    multiplayer online role-playing
    games; Second Life; play
Gardner, R. O., 234n. 14
Gasson, S., 236n. 44
Gavin, J., 47, 98, 238n. 24
Gee, J. P., 248n. 3, 251n. 101
Gefen, D., 239n. 2, 240n. 23, 241n.
    46, 241n. 48
Gemeinschaft, 6
Gender. *See* sociomental connections
    and gender
Gergen, K. J., 238n. 27
Geser, H., 95, 136, 234n. 17, 238n.
    40, 239n. 43, 239n. 45, 241n.
    57, 242n. 3, 243n. 7, 243n. 11,
    243n. 13, 243n. 16, 243n. 18,
    243n. 19, 243n. 22, 245n. 10,
    245n. 12, 245n. 16, 245n. 17,
    245n. 18, 246n. 34, 246n. 46,
    246n. 58, 247n. 63, 247n. 64,
    247n. 5, 247n. 7, 248n. 18,
    248n. 19, 250n. 70, 250n. 72,

250n. 84, 251n. 100, 251n. 108,
    252n. 30
Gibbs, J., 243n. 31, 244n. 34, 251n.
    95, 251n. 97
Giddens, A., 45–46, 139, 141, 238n.
    12, 238n. 41, 247n. 73, 247n. 1
Gilding, M., 239n. 42, 239n. 45
Gillard, P., 251n. 100
Gladwell, M., 236n. 37, 236n. 40
Glaser, J., 250n. 75, 252n. 29
Glasser, T. L., 239n. 2, 239n. 3
Gleason, M., 234n. 25
Global issues, 11, 31, 43, 156, 172,
    184–186, 191–202
Global positioning technology (GPS).
    See surveillance, electronic;
    privacy
Goffman, E., 141, 159, 177, 237n.
    70, 240n. 9, 247n. 1, 248n. 1,
    251n. 99
Goggin, G., 250n. 66
Goleman, D., 235n. 26, 237n.72
Goode, E., 252n. 36
Goodman, S. E., 252n. 9
Goodwin, J., 253n. 48
Google, 52, 132
Gossip and gossiping, 63–64, 73–86,
    115, 187, 217
Gould, P., 235n. 15
Graf, D., 116, 245n. 4
Granovetter, M., 109, 234n. 13,
    244n. 63
Grantham, C., 199
Gray, T., 237n. 64, 242n. 3, 245n. 4
Graya, N., 248n. 17
Green, M., 234n. 25, 239n. 56, 250n.
    75
Gregg, M., 239n. 55, 250n. 81
Greig, M., 252n. 11
Griffin, D. W., 238n. 21
Griffiths, M., 79, 239n. 45, 242n. 78,
    243n. 13, 243n. 31, 243n. 333,
    245n. 13
Grinter, R. E., 233n. 6, 243n. 20,
    243n. 23, 247n. 13, 249n. 27
Gross, N., 238n. 41

Grossman, L., 4, 198, 233n. 5, 253n. 65, 253n. 66
Grotevant, H. D., 249n. 42
Guillen, M., 251n. 1
Guldberg, K., 244n. 50, 244n. 52
Gulia, M., 238n. 13, 244n. 61
Guo, B., 250n. 68

Habermas, J., 237n. 68
Hackers and hacking, 64, 191
Haddon, L., 116, 245n. 8
Hafner, K., 233n. 6
Haisken-Denew, J. P., 234n. 25
Hall, S., 252n. 36
Hale, J. L., 238n. 14
Hallowell, E., 247n. 16
Hamman, R., 234n. 25
Hampton, K. N., 87, 92, 234n. 13, 234n. 25, 237n. 8, 242n. 2, 242n. 6, 244n. 61, 247n. 5, 248n. 21, 250n. 73, 253n. 70, 253n. 71
Hancock, J. T., 245n. 4, 247n. 13
Handelman, D., 240n. 9
Hands, J., 253n. 59
Hanging out in online and mobile spaces, 63, 66, 69–73, 77, 78, 86, 163, 174, 198, 212
Hannigan, S. L., 237n. 67
Hanus, J., 78, 84, 242n. 70, 242n. 71, 242n. 83, 242n. 84
Happiness. See moods
Harassment and technology use, 60, 75, 173, 186–190, 198
Hardey, M., 237n. 2, 238n. 32, 238n. 34, 239n. 43, 239n. 44, 241n. 65, 251n. 97
Hargadon, A., 240n. 34, 240n. 37
Hargittai, E., 138, 243n. 14, 247n. 69, 252n. 10
Haythornthwaite, C., 38, 109, 201, 234n. 13, 237n. 60, 237n. 8, 239n. 46, 242n. 81, 244n. 48, 244n. 61, 245n. 64, 245n. 65, 251n. 108, 253n. 46, 254n. 77
Heer, J., 176, 250n. 83, 250n. 85

Heino, R., 243n. 31
Helsper, E., 248n. 16
Henderson, S., 239n. 42, 239n. 45
Hennessey, M., 238n. 41
Heritage, J., 235n. 18
Herring, S., 253n. 38
Heurtin, J. P., 239n. 43, 241n. 57, 246n. 34, 247n. 62
Hewitt, B., 253n. 39
Hewson, C., 254n. 7, 254n. 9
Hian, L. B., 238n. 9, 238n. 14, 238n. 21
Hiltz, S. R., 245n. 25, 245n. 26, 245n. 27, 246n. 28
Himeno, K., 237n. 65, 242n. 3
Hine, T., 252n. 34
History. See stories
Hitlin, P., 243n. 9
Hoffman, D. L., 252n. 18
Hollander, P., 243n. 32
Holden, D., 236n. 43
Holmes, J. G., 238n. 21
Hossain, L., 239n. 43
Howard, P. N., 252n. 7, 252n. 19
Hu, Y., 238n. 9, 238n. 34, 248n. 20
Huang, J., 250n. 68
Huffaker, D. A., 234n. 16, 248n. 2, 248n. 10, 248n. 15, 249n. 30, 250n. 53, 251n. 90, 251n. 105, 251n. 114
Huizinga, J., 64, 239n. 3, 240n. 7, 240n. 13, 240n. 21
Hulme, M., 246n. 32, 248n. 11, 249n. 32
Humor, 39, 53, 56, 60, 63–64, 73–86, 96, 97, 156
Huse, D., 203
Hyperlinking, 107, 168, 178–182, 198. See also identity, hyperlinking of
Hyperpersonal relationships, 47

Identity
  collective identity, 7, 19, 23–25, 32–36, 38, 44, 74, 97, 104–105, 191–192

Identity *(continued)*
  hyperlinking of, 160, 178–182, 199
  self and individual identity, 13,
    32–36, 38, 63–64, 108, 159–182,
    183, 190, 202, 212
  youth and, 159–166
Identity theft, 54, 186
Igarashi, T., 238n. 39
Igbaria, M., 244n. 58, 252n. 7
IM, instant messages, and instant
    messaging, 3, 5, 8–9, 11, 20,
    29, 39, 43, 56, 59, 69–71, 75,
    77, 80, 88, 90–91, 94, 96–98,
    113, 115, 117, 123, 135, 139,
    142–158, 163, 165, 169–170,
    177, 196, 212, 215–231
Impression management, 159
Industrialization, effect of, 62, 75, 157
Infidelity. *See* cybercheating
Information and information sharing,
    25, 50, 55, 57, 77, 81, 83,
    87–88, 93, 97, 101–104, 107–
    109, 113, 127–128, 139, 147,
    161, 167, 172, 175, 184–186,
    191–193, 215, 217–218
Intersubjectivity, 25–26. *See also*
  common ground
Intimacy and technology use, 5,
    26–31, 37, 39, 43–62, 55, 56,
    74, 77, 81, 88, 91, 97–110, 139,
    145, 153–155, 166, 176–177,
    188, 201, 206, 209
Ishii, K., 243n. 8
Isolation, 69, 106, 117–130, 136,
    157, 171–173, 199–200
Israel, B., 116, 134, 245n. 9, 246n.
    57, 249n. 26
Ito, M., 117, 237n. 64, 237n. 65,
    242n. 3, 245n. 4, 245n. 5, 245n.
    11
Ivory, J., 240n. 16

Jackendoff, R., 235n. 15
Jackson, L. A., 234n. 25, 252n. 18,
    253n. 71

James, W., 40, 237n. 69
Jenkins, H., 246n. 47
Jewkes, Y., 252n. 21
Johnson, S., 67, 68, 240n. 28, 240n.
    29, 240n. 35
Jokes and joking. *See* humor; play
Jones, S., 233n. 4, 236n. 52, 240n.
    17, 240n. 18, 240n. 25
Jordan, T., 23, 184, 235n. 12, 235n.
    13, 235n. 14, 252n. 3, 252n.
    4, 252n. 5, 252n. 6, 252n. 11,
    252n. 13, 253n. 44
Juul, J., 237n. 69, 240n. 13, 240n.
    21, 240n. 26, 240n. 27, 240n.
    37

Kahn, R., 195, 253n. 53, 253n. 57
Kang, H., 252n. 15
Karbo, K., 237n. 5, 237n. 7
Katz, E., 236n. 28
Katz, J., 233n. 13, 234n. 17, 236n.
    28, 238n. 18, 243n. 16, 245n. 1,
    245n. 16, 245n. 23, 246n. 29,
    248n. 13, 248n. 14, 249n. 32,
    251n. 93, 252n. 19, 252n. 25
Kauffmann, R., 244n. 59
Kavada, A., 253n. 61
Kavanaugh, A., 233n. 13, 234n. 25,
    242n. 6, 253n. 46, 253n. 71
Kayahara, J., 238n. 38, 242n. 6
Kellner, H., 117, 195, 245n. 14,
    253n. 53, 253n. 57
Kelly, K., 12, 174, 198, 234n. 27,
    239n. 45, 244n. 49, 250n. 77,
    253n. 64
Kendall, L., 179, 234n. 18, 238n. 11,
    238n. 34, 238n. 37, 240n. 21,
    240n. 39, 241n. 59, 247n. 14,
    250n. 56, 250n. 84, 250n. 89,
    251n. 104, 251n. 106
Kharif, O., 234n. 17
Khazan, O., 240n. 12, 240n. 21,
    240n. 24, 241n. 68, 241n. 69,
    242n. 70, 242n. 73, 242n. 75,
    242n. 80

Kibby, M., 251n. 90
Kidd, D., 253n. 42
Kiesler, S., 252n. 5
Klastrup, L., 21, 235n. 9, 235n. 10,
    235n. 11, 240n. 21
Kling, R., 244n. 51
Kornet, A., 248n. 22, 248n. 24
Koyuncu, C., 244n. 58
Krcmar, M., 240n. 21
Krueger, B., 246n. 47, 253n. 47
Kulikova, S. V., 253n. 56
Kumar, R., 243n. 8
Kyte, 116

Lamb, B., 234n. 16
Langman, L., 195, 253n. 58
Larose, R., 239n. 55, 239n. 57, 242n.
    70, 242n. 72, 242n. 74, 242n.
    76
Lawson, H., 243n. 31, 244n. 34
Lea, M., 234n. 13, 238n. 18
Leach, E., 235n. 15
Learning. See distance learning;
    information; education; youth
    and learning
Leck, K., 243n. 31, 244n. 34
Lederer, D. A., 244n. 39
Lee, H., 239n. 58, 239n. 61, 250n.
    76
Legge, K., 233n. 6, 245n. 20, 246n.
    31, 249n. 32, 249n. 35, 252n.
    25, 252n. 26
Lemann, N., 253n. 43, 253n. 50
Lengel, L., 238n. 18
Lenhart, A., 32, 234n. 16, 236n. 38,
    236n. 39, 241n. 40, 243n. 9,
    243n. 23, 243n. 26, 243n. 30,
    247n. 66, 247n. 9, 247n. 10,
    248n. 15, 249n. 26, 249n. 30,
    249n. 32, 249n. 36, 249n. 45,
    249n. 48, 250n. 80, 251n. 92,
    251n. 109, 252n. 31
Letters, 17, 22, 149, 153, 155, 176,
    188, 212, 226
Levinson, P., 252n. 8, 252n. 30

Lewin, K., 235n. 15
Lewis, O., 243n. 9
Licoppe, C., 37, 237n. 59, 237n. 65,
    239n. 43, 241n. 57, 246n. 34,
    247n. 62
Like-mindedness. See common
    ground; cognitive resonance;
    intersubjectivity
Lin, D., 250n. 61
Lin, N., 233n. 13, 250n. 53
Lindsay, S., 239n. 56, 252n. 11
Linenthal, E. T., 236n. 55
Ling, R., 92, 121, 133, 163, 234n.
    14, 237n. 63, 238n. 18, 238n.
    31, 239n. 58, 241n. 40, 241n.
    44, 241n. 60, 242n. 3, 243n. 13,
    243n. 16, 243n. 18, 243n. 20,
    243n. 22, 243n. 25, 245n. 16,
    245n. 20, 245n. 23, 245n. 24,
    245n. 25, 245n. 26, 245n. 27,
    246n. 28, 246n. 29, 246n. 37,
    246n. 53, 246n. 55, 247n. 70,
    248n. 13, 248n. 16, 249n. 26,
    249n. 28, 249n. 32, 249n. 33,
    250n. 56, 251n. 93, 251n. 94,
    252n. 8, 252n. 25, 252n. 26,
    252n. 30, 252n. 31
Linguistic style, online, 19, 21, 28,
    30, 31, 46–48, 57, 76, 162, 163–
    165, 168–169, 177, 184, 197
Linn, M. C., 244n. 42
Liu, S. B., 245n. 25, 245n. 26, 245n.
    27, 246n. 28
LiveJournal, 234n. 16
Livingstone, S., 248n. 10, 248n. 16,
    248n. 18
Lo, V. H., 234n. 18, 237n. 63, 242n.
    3, 250n. 72
Loftus, E., 237n. 67
Lombard, M., 236n. 28, 237n. 58
Lomrantz, T., 242n. 70, 249n. 38
Longino, C., 234n. 14
Long, K., 242n. 70
Loneliness. See isolation
Losh, S., 252n. 11

Love and romance, 10, 12, 17, 26–28, 43–51, 59, 72, 76, 81, 87–91, 97–110, 122, 130, 137, 148, 172, 183, 198, 221, 223, 231. *See also* flirting; sexual activity
Luckmann, T., 40, 141, 237n. 68, 237n. 69, 237n. 70, 247n. 1
Lurking, 72, 173–178, 220, 226–231
Lyon, D., 246n. 47
Lytras, M., 243n. 15

MacKinnon, G., 244n. 50
Madden, M., 234n. 16, 241n. 40, 243n. 9, 243n. 26, 243n. 30, 244n. 53, 244n. 58, 247n. 66, 247n. 9, 248n. 15, 249n. 30, 249n. 36, 251n. 92, 251n. 109
Madge, C., 252n. 12
Magid, L., 246n. 40, 246n. 42, 252n. 26
Mann, C., 254n. 4, 254n. 11, 254n. 12, 254n. 13, 254n. 14
Martin, S. P., 252n. 10
Marx, G. T., 234n. 20
Mason, D., 247n. 71, 247n. 72
Mason-Schrock, D., 236n. 43
Mass media, 4, 20–22, 25, 33, 161–162, 165, 188, 191–193, 196, 200. *See also* radio; television
Massaneri, A., 236n. 48, 253n. 63
Massive multiplayer online role-playing games (MMOs, MMORPGs), 21, 66, 78, 80, 217. *See also* games; Second Life; play
Matsuba, M., 248n. 19, 249n. 40, 250n. 71
McCaughey, M., 196, 247n. 72, 253n. 41, 253n. 43, 253n. 60
McCullagh, D., 246n. 49, 246n. 50
McGowan, K., 235n. 25
McKenna, K. Y. A., 46, 48, 144, 234n. 25, 235n. 24, 238n. 15, 238n. 27, 238n. 28, 238n. 32, 238n. 33, 239n. 45, 240n. 18, 243n. 31, 243n. 33, 244n. 36, 247n. 6, 249n. 44, 250n. 59, 250n. 62, 250n. 72, 251n. 96

McLoughlin, I. P., 243n. 27, 252n. 5
McMillan, S., 248n. 16, 249n. 38
McQuaid, R., 252n. 11, 252n. 18
Mead, G. H., 235n. 7, 240n. 6, 248n. 4, 248n. 6, 248n. 7
Medical issues. *See* social support online and medical issues
Meho, L. I., 254n. 3, 254n. 5
Mehra, B., 250n. 61
Memorialization online. *See* web memorials
Memory, collective. *See* collective memory
Mental health and illness, 40–41, 64, 97, 128–130, 171. *See also* anxiety; moods
Mental infrastructure. *See* cognitive infrastructure of society
Mental maps, 23–25, 35
Merkel, C., 250n. 61
Merton, R., 108, 244n. 60
Mesch, G., 249n. 40
Message boards. *See* discussion boards
Metzger, M. J., 236n. 45, 239n. 42, 244n. 58, 245n. 15, 247n. 72
Meyrowitz, J., 25, 235n. 16, 235n. 17
Micro-coordination of activities, 88, 95, 129, 156
Microsoft, 132
Mikula, M., 236n. 50
Millar, J., vi, 204
Mills, E., 246n. 49
Mitra, A., 236n. 37, 250n. 61
Miyaki, Y., 252n. 31
Miyata, K., 234n. 17, 242n. 3, 249n. 27, 249n. 28, 249n. 39, 252n. 31
MMO, MMORPG. *See* massive multiplayer online role-playing games
Monk, A., 252n. 30
Moods, 56–62, 68–69, 84, 91, 100, 126–130, 148, 212

Nakajima, I., 242n. 3
Nakamura, L., 237n. 65, 250n. 55
Nassef, Y., 252n. 9
Nastri, J., 243n. 23, 245n. 4, 247n. 13, 251n. 92

Nation and issues of nationalism, 4, 8, 18, 20, 29, 51, 184–186, 191–198
Navarro, V., 131, 245n. 17, 246n. 41
Neimark, J., 237n. 67
Networks and networking. See social networks and networking
Networked individualism, 92
Newby, H., 233n. 12
Newell, C., 250n. 66
Newsweek online, 238n. 16
Nippert Eng, C., 76, 241n. 67, 246n. 58, 248n. 23
Nonnecke, B., 241n. 46, 241n. 47, 241n. 48, 250n. 78
Nordland, R., 246n. 47
Norrick, N. R., 241n. 54
Novak, J., 252n. 18
Nowak, K. L., 238n. 9, 240n. 21

O'Brien, J., 235n. 9, 250n. 55, 251n. 103
O'Connor, H., 252n. 12
O'Harrow, R., 246n. 38, 246n. 47, 246n. 54
Okabe, D., 117, 236n. 41, 237n. 64, 237n. 65, 242n. 3, 245n. 4, 245n. 5, 245n. 11, 245n. 12
Oksman, V., 245n. 23
Oldenburg, R., 69, 241n. 41, 241n. 43
Online research, 205–213
Ontological security, 139–140
Open source software, 4
Organic solidarity, 158

Page, S., 246n. 49
Palen, L., 241n. 57, 245n. 17, 245n. 22, 245n. 25, 245n. 26, 245n. 27, 246n. 28
Palmer, J., 241n. 54
Pan, B., 239n. 43
Parker, S., 247n. 17
Parks, M. R., 238n. 38, 250n. 86
Patterson, S., 234n. 25, 242n. 6, 253n. 46, 253n. 71
Paul, P., 242n. 70, 242n. 83

Pavlik, J., 236n. 36
Pecchioni, L. L., 239n. 54
Pedophilia, 186
Pena, J., 245n. 4, 247n. 13
Perlmutter, D. D., 253n. 56
Persichitte, K., 206, 254n. 1, 254n. 2, 254n. 6, 254n. 7, 254n. 8
Peter, J., 243n. 28, 247n. 4, 248n. 19, 248n. 20
Peters, S., 246n. 32, 248n. 11, 249n. 32
Petronio, S., 241n. 67
Photos and photo sharing, 4, 9, 11, 18, 19, 31–33, 35–36, 38, 46, 58, 72, 90, 98, 99, 108, 115, 117, 132, 149, 161, 163, 175–176, 188
Physical discomfort and fatigue in technology use, 61, 64, 81, 212
Pickard, V., 253n. 45
Pilkington, R., 244n. 50, 244n. 52
Pischner, R., 234n. 25
Pisello, T., 246n. 33
Pitts, V., 252n. 12
Plant, S., 169, 250n. 54
Play, 5, 13, 63–86, 109, 169. *See also* games; hanging out
Podcasts, 4, 8, 10, 20, 31, 47, 69
Political issues, 30, 51, 60, 183–202
Political participation, 13, 20, 105, 189–199, 191–198, 229
Pornography, 78, 84–85, 161, 174, 186
Porter, C. E., 244n. 58
Portes, A., 233n. 13
Portfolio of social connections, 31, 93, 109, 155, 180
Poster, M., 236n. 36
Prager, K. J., 238n. 14
Preece, J., 233n. 3, 233n. 11, 234n. 13, 234n. 18, 235n. 19, 235n. 23, 238n. 18, 239n. 48, 239n. 58, 239n. 59, 239n. 60, 241n. 46, 241n. 47, 241n. 48, 241n. 52, 244n. 62, 250n. 56, 250n. 72, 250n. 78, 251n. 95

Presence in online and mobile spaces, 18, 37–42, 69–73, 113–140, 157, 173–179. *See also* proximity; ambient copresence
Princess Diana, 35
Pringle, D., 242n. 70, 246n. 47
Privacy concerns, 73, 76, 77, 114, 130–140, 157, 175, 186, 210–211. *See also* surveillance, electronic
Procrastination, 69, 82–84, 209
Proximity in online and mobile spaces, 18, 37–42, 113–140. *See also* presence
Psychology Today, 235n. 25
Pure relationships, 45–46
Puro, J. P., 246n. 34
Putnam, R., 233n. 13, 253n. 69

Qualitative research, 205–213
Quan-Haase, A., 237n. 64, 242n. 3, 243n. 23, 245n. 4, 247n. 17
Quasi-synchronicity. *See* synchronicity and quasi-synchronicity

Raab, C., 247n. 71, 247n. 72
Race. *See* sociomental connections and race
Radcliffe, A. M., 239n. 57
Radin, P., 53, 239n. 50, 239n. 51, 239n. 52
Radio, 3, 10, 20, 29, 193. *See also* mass media; podcast
Rainie, L., 241n. 45, 242n. 73, 243n. 9, 244n. 44, 244n. 53, 244n. 55, 245n. 7, 246n. 58, 246n. 60, 249n. 27, 249n. 36, 251n. 109, 251n. 110, 252n. 31, 253n. 68
Rakow, L., 131, 245n. 17, 246n. 41
Ramirez, A., 250n. 87
Rao, H., 252n. 15
Rautianen, P., 245n. 23
Reality and "realness" of sociomental phenomena, 3–4, 10, 35, 37–43, 48, 61, 64, 169–173, 177, 180–183, 190, 201, 243n. 21, 243n. 22

Reid, D., 236n. 43, 238n. 13, 242n. 3, 245n. 10, 247n. 8, 249n. 28, 249n. 51, 250n. 58, 250n. 72
Reid, F., 242n. 3, 245n. 10, 247n. 8, 249n. 28, 249n. 51, 250n. 58, 250n. 72
Reinitz, M. T., 237n. 67
Religion, 18–20, 40, 60, 192
Remote mothering, 131
Renninger, K. A., 244n. 48, 244n. 51
Rheingold, H., 10, 234n. 19, 234n. 25, 243n. 11, 251n. 108, 253n. 71
Rice, R. E., 233n. 13, 234n. 17, 252n. 18, 252n. 25
Richardson, J., 237n. 58
Richtel, M., 245n. 5
Ridings, C., 239n. 2, 240n. 23, 241n. 46, 241n. 48
Ring tones, 144, 163, 177
Rituals, 17, 19–21, 31, 60
Riva, G., 245n. 1
Robbins, K., 252n. 8
Roberts, P., 236n. 52
Robinson, J. P., 252n. 10
Rogers, C., 250n. 57
Roloff, M. E., 245n. 2
Romance. *See* love and romance
Rosen, J., 246n. 38, 247n. 66, 247n. 67
Ruane, J., 236n. 28, 236n. 54, 237n. 7, 238n. 13
Ryan, R., 180, 181

Sadness. *See* moods
Safety, 73, 76, 113, 120–130, 133, 135–136, 139, 161–162, 164, 188, 190. *See also* emergencies
Sakkopoulos, E., 243n. 15
Salzman, M., 241n. 57
Sanders, C., 242n. 72
Sanders-Jackson, C., 243n. 9
Sandvig, C., 239n. 2, 239n. 3, 239n. 4, 240n. 5, 240n. 14
Sassen, S., 244n. 54
Schmidt, K., 244n. 54
Schneider, S. M., 236n. 52, 237n. 56

Schopler, J. H., 244n. 54
Schouten, A., 248n. 19
Schramm, W., 26, 235n. 22
Schrock, D., 171, 236n. 43, 250n.
    63, 250n. 64, 251n. 103, 253n.
    49
Schumacher, P., 250n. 71
Schutz, A., 26, 40, 235n. 18, 235n.
    22, 237n. 69
Schwartz, J., 233n. 4
Search engines and searching, 103,
    139
Second Life, 4, 66, 70
Secrets, 54, 77, 129, 156–157, 167,
    169, 177
Self expression online, 13, 41, 149,
    159–182, 190, 199, 209
Sexual activity and technology use, 40,
    76, 78–86, 85, 97–110, 202. See
    also cybersex; love and romance;
    flirting
Sexual orientation. See sociomental
    connections and sexual orientation
Shafer, S., 252n. 10
Shah, D., 253n. 50
Shared knowledge. See common
    ground
Sheller, M., 136, 246n. 58, 247n. 65
Shibutani, T., 237n. 69, 248n. 9
Shopping. See commerce
Short, J., 237n. 57
Shumar, W., 244n. 48, 244n. 51
Sidner, C., 246n. 33
Siegl, E., 236n. 54
Silver, D., 105, 196, 236n. 48, 244n.
    56, 250n. 55, 253n. 54, 253n.
    62, 253n. 63
Simmel, G., 73, 179, 233n. 8, 234n.
    1, 236n. 51, 241n. 51, 241n. 67,
    251n. 108
Simmons, S., 238n. 41
Skovholt, K., 243n. 23, 244n. 54
Slovacek, C., 235n. 10
Smallwood, A. M. K., 243n. 9
Smircich, L., 236n. 44
Smyth, K., 243n. 13
Social capital, 7, 53, 109, 184–186

Social control in technology use. See
    control, social
Social networks and networking, 4–5,
    7–8, 10–11, 24–26, 31, 49, 63,
    66, 69, 71, 75, 85, 87–110, 113,
    116, 119–120, 137, 138, 142,
    163, 165, 173, 175, 184, 189,
    190, 191–197, 192–193, 195–
    196, 200–201, 215–231. See also
    networked individualism
Social movements, 10, 191. See also
    political participation
Social presence. See presence; proximity
Social status, 53, 75, 108, 126, 127,
    155, 184, 197
Social support online, 10, 12, 39, 44,
    51–56, 62, 87, 102–104, 113,
    120, 171–173, 228–229
    and medical issues, 30, 51, 53–56,
    58, 172–173, 175, 228–229
Socialization, 79, 160–166
Sociomental connections, 10–11,
    17–42
    and age, 47, 160, 169–171, 173,
    179, 212, 220
    and disability, 171–173, 185
    and gender, 12, 30, 48, 65, 75,
    121, 126–127, 160–161, 169–
    171, 177, 179, 181, 185, 212
    and race, 12, 30, 160, 169–171,
    179, 181, 185, 188, 213
    and sexual orientation, 52, 160,
    169–171, 179, 181
Sociomental space, 18, 22–25, 31, 35,
    37, 40, 42, 50, 88, 106, 124,
    134, 172, 174, 179, 187, 200,
    205
Solomon, D. H., 245n. 2
Solove, D. J., 246n. 35, 246n. 38,
    246n. 47
Spam, 126, 128
Sparks, L., 239n. 54
Spears, R., 234n. 13, 238n. 18
Spitzberg, B. H., 242n. 77, 250n. 71
Spontaneity in interaction, 73, 95–96,
    142, 155–158, 171, 193–195,
    206

Sports, 19, 20, 27, 63, 91, 95, 188
Sproull, L., 252n. 5
Spyware, 128
Stalking, 186. *See also* harassment;
 bullying
Status. *See* social status
Stea, D., 235n. 15
Steinkuehler, C., 69, 240n. 21, 240n.
 39, 241n. 41, 241n. 42, 241n.
 49
Stephenson, W., 239n. 2
Sterling, B., 22–23, 235n. 12, 235n.
 13, 253n. 44
Stewart, F., 254n. 4, 254n. 11, 254n.
 12, 254n. 13, 254n. 14
Stigma of sociomental connecting,
 3–4, 30–31, 40–42
Stoller, E., 234n. 14
Stone, L., 71, 116, 245n. 5, 245n. 7,
 246n. 60, 251n. 101
Stories, 31–36, 39, 42, 66, 74, 154,
 161, 168, 175–176, 182, 193,
 207
Strategies for interaction, 145,
 173–182
Stress. *See* anxiety; mental health and
 illness
Strogatz, S., 235n. 26
Suarez, S., 251n. 2
Suellentrop, C., 68, 240n. 7, 240n.
 30, 240n. 31, 240n. 32, 240n.
 33, 251n. 109
Sugiyama, S., 238n. 18, 248n. 13,
 248n. 14, 249n. 32, 251n. 93
Suhail, K., 242n. 70, 242n. 76
Suler, J., 48, 238n. 25, 238n. 26,
 239n. 45, 249n. 43
Surveillance, electronic, 130–140, 186,
 190, 198. *See also* privacy
 concerns
Sveningsson, M., 238n. 34
Svennevig, J., 243n. 23, 244n. 54
Swan, K., 237n. 58
Symbols and symbolic expressions, 17,
 18, 84, 121, 126, 127, 128, 155,
 163–164, 169, 177

Synchronicity and quasi-synchronicity, 25–31, 98, 142–145, 152,
 194–195, 205–206. *See also* temporal symmetry

Takai, J., 238n. 39
Talmund, I., 249n. 40
Tan, H., 240n. 8
Tannen, D., 250n. 56
Tanner, E., 236n. 46, 236n. 50,
 237n. 1
Technological determinism, 82, 85,
 188
Technological dowry, 131
Television, 20, 27, 28, 29, 36, 60, 70,
 83, 87, 107, 134, 165, 174, 200,
 226, 230. *See also* mass media
Temporal symmetry, 17, 29–30, 68,
 142. *See also* synchronicity
"Tethered" to technology, 5–6, 113,
 119–120, 133–137, 156
Text circles, 8, 20, 106, 147, 227
Texting and text messaging, 3, 5,
 8–11, 20, 29, 39, 43, 50, 54,
 56–59, 63, 69, 75, 76, 91–92,
 94, 96, 97, 115, 117, 118, 125,
 132, 139, 142–158, 163, 169,
 170, 173, 177, 188, 194, 212,
 215–231
Tharp, D. D., 254n. 1
Thomas, A., 161, 181, 234n. 21,
 248n. 3, 248n. 10, 248n. 19,
 249n. 21, 251n. 109, 251n. 111
Thomas, D. S., 254n. 76
Thomas, W. I., 10, 234n. 21, 254n.
 76
Thompson, K., 252n. 32
Thoreau, E., 236n. 36, 250n. 66,
 250n. 67, 252n. 12
Thornburgh, N., 242n. 82
Thurlow, C., 46, 238n. 18, 238n.
 19, 238n. 20, 238n. 23, 238n.
 25, 240n. 18, 241n. 66, 243n.
 27, 243n. 28, 244n. 35, 251n.
 1, 252n. 3, 252n. 7, 252n. 10,
 252n. 11, 252n. 17, 252n. 17

Tidwell, L. C., 235n. 10, 251n. 94
Tilly, C., 236n. 37, 236n. 42
Timms, D., 252n. 14
Tomic, A., 238n. 18
Tonnies, F., 6, 233n. 11
Tosca, S., 21, 235n. 9, 235n. 10,
    240n. 21
Tracey, K., 254n. 74
Tranter, B., 252n. 16
TripAdvisor, 108
Trolling, 59–60, 187–188
Trust and technology use, 44, 51–56,
    62, 107, 137, 138–140, 168,
    177, 192
Tsakalidis, A., 243n. 15
Tuffin, K., 250n. 65
Tugend, A., 246n. 41, 246n. 43, 246n.
    44, 247n. 16, 249n. 37, 249n. 38
Turkle, S., 44, 61, 131, 134, 162,
    178, 239n. 62, 240n. 36, 242n.
    79, 246n. 29, 246n. 32, 246n.
    45, 247n. 61, 249n. 41, 250n.
    55, 251n. 101, 251n. 102
Turner, M. A., 252n. 8, 253n. 72
Turow, J., 233n. 4, 238n. 41
Twitter, 39, 116
Typified others, 20

Urbina, I., 236n. 53
Utz, S., 240n. 23

Valkenburgh, P., 243n. 28, 247n. 4,
    248n. 19, 248n. 20
Van Dijk, J. A., 101, 244n. 43, 251n.
    109, 251n. 1, 252n. 16, 252n.
    17, 252n. 19
van't Hooft, M., 244n. 49
Van Valey, T., 248n. 16
Venkatesh, M., 234n. 13
Venolia, G., 246n. 33
Vidcasts, 8, 47, 90, 149. See also
    vodcasts; podcasts
Videl, L. A., 236n. 52
Video sharing. See also vidcasts; vod-
    casts; camcorders), 4, 8, 10, 17,
    32, 38, 47, 84, 90

Vilhjalmsson, H. H., 234n. 16
Violence and technology, 85, 187. See
    also harassment
Virnoche, M. E., 234n. 20
Viruses, 128
Vodcasts, 8, 10, 31, 38, 47, 90, 149.
    See also vidcasts; podcasts
Voyeurism, 138, 159–182
Vroman, A., 252n. 9, 253n. 46

Wade, M., 68, 240n. 30, 240n. 35,
    251n. 109
Wagner, G., 234n. 25, 244n. 47,
    246n. 47, 253n. 71
Wallace, P., 47, 234n. 13, 234n. 18
Walther, J., 47, 235n. 9, 235n. 10,
    238n. 9, 238n. 18, 238n. 21,
    238n. 23, 238n. 38, 250n. 86,
    251n. 91, 251n. 94, 253n. 72
Warnick, B., 236n. 52, 237n. 56
Warren, R. L., 234n. 13
Wasko, M. M., 239n. 2
Watt, J., 238n. 9
Watts, D., 175
"We feeling," 26–27
Webcams. See also video sharing;
    amcorders), 32, 46, 47, 88, 90,
    99
Web memorials, 34–36, 46
Wei, R., 234n. 18, 237n. 63, 242n. 3,
    250n. 72
Weinberg, N., 243n. 13
Weinberger, D., 116, 245n. 6
Wellman, B., 87, 92, 137, 139, 201,
    233n. 7, 234n. 13, 234n. 25,
    237n. 64, 237n.8, 238n. 13,
    238n. 38, 242n. 81, 242n. 2,
    242n. 3, 242n. 6, 243n. 11,
    243n. 12, 243n. 23, 243n. 24,
    244n. 53, 244n. 61, 245n. 4,
    247n. 5, 247n. 17, 248n. 21,
    250n. 73, 251n. 108, 253n. 70,
    253n. 71, 254n. 75, 254n. 77,
    254n. 78
Wenger, E., 244n. 42
Weslander, E., 251n. 113

White, R., 235n. 15
Whittaker, S., 246n. 33
Whittemore, G., 241n. 50
Whitty, M., 241n. 62, 241n. 63,
    241n. 64, 243n. 27, 244n. 38,
    244n. 340
Wigand, R. T., 239n. 43
Wiggins, J. D., 244n. 39
Wikipedia, 8, 13
Wikis, 8, 31, 103, 217
Willard, N., 249n. 38, 252n. 20,
    252n. 26
Williams, D., 66, 69, 79, 240n. 21,
    240n. 39, 241n. 41, 241n. 42,
    241n. 49, 243n. 17
Williams, E., 237n. 57
Williams, L., 246n. 44
Williams, M., 253n. 38
Williams, P., 244n. 50
Williams, R., 80, 242n. 70, 242n. 80
Williams, S., 246n. 44
Williamson, K., 250n. 66, 250n. 69
Willis, S., 252n. 16
Wilson, B., 181, 240n. 8, 251n. 112
Winnicott, D. W., 237n. 73
Wired magazine, 12
Wood, R. T., 80, 242n. 70, 242n. 80,
    244n. 59
Work and the workplace, 25, 44, 56,
    57, 67–73, 81, 86, 87, 91, 93,
    94, 101–110, 118, 119, 126,
    132–135, 140, 145, 150, 152,
    176, 178–180, 181, 183–186,
    198, 202, 212, 215–231. See also
    colleagues

Yahoo, 67, 132
Yee, N., 79, 80, 240n. 19, 242n. 80
Young, K. S., 241n. 57, 244n. 38,
    244n. 39, 244n. 48, 254n. 1
Youth, 67–68, 127–128, 132–135,
    140
  and identity, 159–166, 178–182
  and learning, 83–84, 161–163
  and play, 40, 63–86
  and technology use, 39, 64, 69–73,
    78–86, 97, 102–105, 115–120,
    130–131, 134, 149, 160–166,
    176–177, 178–182, 199–202,
    212
YouTube, 10. See also video sharing
Yoshida, T., 238n. 39
Yoshii, H., 237n. 65, 242n. 3
Yttri, B., 243n. 20, 243n. 22
Yu, H., 253n. 51

Zerubavel, E., 29, 31, 234n. 22,
    236n. 28, 236n. 29, 236n. 34,
    236n. 35, 237n. 70, 251n. 107,
    251n. 114
Zhao, S., 243n. 10, 245n. 1
Zimmerman, E., 239n. 4, 240n. 13,
    240n. 20